Physiotherapy: Foundations for Practice

Key Issues in Musculoskeletal Physiotherapy

Titles in the series

Key Issues in Cardiorespiratory Physiotherapy
Edited by Elizabeth Ellis & Jennifer Alison

Key Issues in Musculoskeletal Physiotherapy
Edited by Jack Crosbie & Jenny McConnell

Key Issues in Neurological Physiotherapy
Edited by Louise Ada & Colleen Canning

Physiotherapy: Foundations for Practice

Series Editors: Janet H. Carr & Roberta B. Shepherd

Key Issues in Musculoskeletal Physiotherapy

Editors:

Jack Crosbie MSc, PhD, Grad. Dip. Phys., Dip.T.P.

Senior Lecturer, School of Physiotherapy, University of Sydney, New South Wales, Australia.

Jenny McConnell B.App.Sc., M.Biomed.E., Grad.Dip. Manip.Ther.

Physiotherapist, Sports Science Research Centre, University of Sydney, New South Wales, Australia

Butterworth-Heinemann Ltd
Linacre House, Jordan Hill, Oxford OX2 8DP

℞ A member of the Reed Elsevier group

OXFORD   LONDON   BOSTON
MUNICH   NEW DELHI   SINGAPORE   SYDNEY
TOKYO   TORONTO   WELLINGTON

First published 1993
Reprinted 1994

© Jack Crosbie and Jenny McConnell 1993

British Library Cataloguing in Publication Data
Key Issues in Musculoskeletal
Physiotherapy. – (Physiotherapy:
Foundations for Practice Series)
  I. Crosbie, Jack   II. McConnell, Jenny
  III. Series
  615.8

ISBN 0 7506 0177 9

Library of Congress Cataloguing in Publication Data
Key issues in musculoskeletal physiotherapy/editors, Jack Crosbie,
  Jenny McConnell.
  p.   cm. – (Physiotherapy)
  Includes bibliographical references and index.
  ISBN 0 7506 0177 9
  1. Musculoskeletal system – Physical therapy.   I. Crosbie, Jack.
  II. McConnell, Jenny.   III. Series.   IV. Series: Physiotherapy
  (Oxford, England)
  [DNLM: 1. Bone Diseases – rehabilitation.   2. Joint Diseases –
  rehabilitation.   3. Muscular Diseases – rehabilitation.   4. Physical
  Therapy.   WE 550 K44 1993]
  RC925.5.K48   1993
  616.7'062–dc20   92–48427
                                                        CIP

THIS EDITION FOR SALE IN
THE REPUBLIC OF THE PHILIPPINES

Printed in Singapore by International Press Co. Pte Ltd

Contents

Contributors

John H. Bland, MD, FACP
Professor of Medicine – Rheumatology, Emeritus, Rheumatology and Clinical Immunology Unit, Department of Medicine, University of Vermont College of Medicine, Burlington, Vermont, USA.

Nikolai Bogduk, BSc(Med), MB, BS, MD, PhD, DipAnat., Hon. FACRM, Hon. MMTAA
Professor of Anatomy, University of Newcastle, New South Wales, Australia.

Robert Cooper, MD, MRCP
Consultant Rheumatologist, Pinderfields General Hospital, Wakefield, United Kingdom.

Jack Crosbie, MSc, PhD, Grad. Dip.Phys., Dip.T.P.
Senior Lecturer, School of Physiotherapy, University of Sydney, New South Wales, Australia.

Rob Herbert, B.App.Sc., M.App.Sc.
Lecturer, School of Physiotherapy, University of Sydney, New South Wales, Australia.

Jenny McConnell, B.App.Sc., M.Biomed.E., Grad. Dip. Manip. Ther.
Physiotherapist, Sports Science Research Centre, University of Sydney, New South Wales, Australia.

Maria Stokes, PhD, MCSP.
Director of Research Services, Royal Hospital and Home, Putney, London, United Kingdom.

Series Editors

JH Carr, DipPhty, EdD, FACP
RB Shepherd, DipPhty, EdD, FACP

The series editors are professors at the School of Physiotherapy, Faculty of Health Sciences, The University of Sydney, Sydney, Australia and have collaborated on several books and articles. They are developing, based on research in the movement sciences, a theoretical framework for the rehabilitation of individuals with movement dysfunction.[1,2] As a result of this theoretical work, they have been invited to lecture and present papers throughout the world. Their individual research activity is in the investigation of normal and disabled performance of everyday motor actions.

1. Carr J.H., Shepherd R.B. (1987). *A Motor Relearning Programme for Stroke*, 2nd edn. Oxford: Butterworth-Heinemann.
2. Carr J.H., Shepherd R.B., eds (1987). *Movement Science Foundations for Physical Therapy in Rehabilitation*. Rockville, Md.: Aspen.

Series Editors' Preface

Modern physiotherapy is in the process of considerable change. Physiotherapists are increasingly turning away from a therapeutic function based principally on a medical diagnosis, to an analytic, diagnostic[1] and therapeutic function, based on the relevant biological science, behavioural science and biomechanics, in which the medical diagnosis is just one of many sources of information relevant to clinical practice. From its original development as an extension of medical practice (hence the term 'paramedical'), physiotherapy is emerging as an independent applied science.

We have proposed elsewhere[2,3] that the broad area of movement science, which encompasses those parts of the biological and behavioural sciences and biomechanics related to human movement, should form the basis from which rehabilitation strategies are built and tested. This view, we believe, will take physiotherapy into the next century as a health science with its own distinct clinical expertise, equipped also to contribute significantly, through experimental investigation, to the understanding of human function.

The role of the modern clinical physiotherapist in the identification and analysis of problems, therapeutic applications and the improvement of functional motor performance must, therefore, increasingly depend upon four factors: 1) keeping up to date with progress in the relevant behavioural and biological sciences, and biomechanics; 2) developing judgement to see and understand what is relevant to clinical practice in such theoretical and data-based information; 3) devising analytic and intervention strategies from this material; and 4) testing whether or not these strategies lead to improvement in some aspect in human function, whether it be physiological or behavioural. The role of the academic and research physiotherapist must increasingly be to investigate clinical observations as part of enriching the data-base on human behaviour.

The purpose of this series of books is to illustrate three processes critical to clinical practice: the deduction of clinical implications from theoretical and data-based material; the development of intervention and measurement strategies for clinical practice; and the testing of outcome.

With this in mind, each volume in the series will include chapters designed to incorporate up-to-date theoretical and data-based information together with illustrations of the use of this information in the identification and analysis of specific clinical problems and in the development and testing of intervention strategies. It should be noted that the individual books are designed to illustrate the *process* out of which clinical practice emerges. Each book will provide, therefore, a selection rather than a complete coverage of what may be considered relevant material.

The series is designed principally for undergraduate physiotherapy students in order to help them acquire the mental skills necessary for becoming a clinician and for qualified physiotherapy clinicians who wish to update their knowledge. Post-graduate students will find questions raised that should stimulate a vigorous search for answers.

JH Carr, RB Shepherd
Series Editors

1. Sahrmann S. (1987). *Diagnoses by physical therapists. Application to trunk imbalance.* Paper presented at the Annual Conference of the Australian Physiotherapy Association, Canberra.
2. Carr J.H., Shepherd R.B. (1987). *A Motor Relearning Programme for Stroke.* 2nd edn. Oxford: Butterworth-Heinemann.
3. Carr J.H., Shepherd R.B., eds. (1987). *Movement Science. Foundations for Physical Therapy in Rehabilitation.* Rockville, Md.: Aspen.

Preface

In contemplating the area of musculoskeletal physiotherapy, one is conscious of the breadth of both the topic itself and the literature related to it. Despite its apparently complex nature, musculoskeletal physiotherapy is, fundamentally, directed towards the application of treatment interventions aimed at three things, either singly or together: 1) the compliance of tissue; 2) the performance characteristics of muscle; and 3) the perception of pain. With this in mind, the physiotherapist needs, above all else, to be familiar with material relevant to normal function and to the pathophysiology, mechanics and psychosociology of musculoskeletal disorders and to be able to apply the material to clinical practice. Ultimately, the physiotherapist should aim to improve the performance of those actions with which the patient has difficulty. The restoration of an optimal motor performance is an indication of the overall effectiveness of the intervention.

The purpose of this book is to provide the undergraduate physiotherapy student with background information on some of the important, one might say essential, topics relating to the treatment of musculoskeletal conditions. This is neither a textbook of orthopaedic conditions, nor an instruction manual in musculoskeletal physiotherapy techniques; numerous other texts cover these areas more thoroughly than this volume has set out to do. Instead, a number of issues, central to physiotherapy intervention, are explored and discussed from the basis of contemporary scientific knowledge. No recipes for treatment are provided here; it is the reader's responsibility to incorporate the material discussed into their own clinical practice.

The pathological basis of musculoskeletal disorders is extensive, encompassing traumatic incidents involving disruption of otherwise healthy tissues, degenerative processes consequential on life-style, congenital structural anomalies or acquired organic disease. It may be this diversity of pathology that has stimulated the development of a vast repertoire of treatment strategies. As mentioned above, however, the problem base in musculoskeletal physiotherapy is actually confined to three elements.

The wide range of skills which the physiotherapist must bring to bear on these problems has evolved over many years and has drawn on many sources. The basis of much of this contemporary physiotherapy practice is an interesting mixture of inductive and deductive reasoning. Many treatment options are derived from observations of positive patient responses to a particular intervention. Others have developed from theoretical principles based on physical or biological laws.

Good clinical practice must be based upon a versatile and flexible approach to the solution of problems. There is, therefore, no inherent invalidity in either the deductive or inductive approach. However, the acceptance of limited anecdotal evidence as indicative of proof of the efficacy of a treatment is entirely inappropriate, as is the belief that the findings of a carefully controlled, but highly specific, experiment can be generalized widely without qualification. The place of scientific enquiry in physiotherapy is assured and its importance is two-fold; firstly, therapists can formulate models that help to explain the mechanisms underlying musculoskeletal dysfunction, and secondly they can apply and evaluate intervention directed at the mechanisms rather than their clinical manifestations. In both instances a combination of theoretical and empirical approaches can be employed. The modern physiotherapist needs to be conscious of the extent to which assumptions concerning intervention are valid and current. The physiotherapist must be prepared to generate testable hypotheses relating to clinical practice and to investigate these using sound scientific principles.

Treatment strategies in musculoskeletal physiotherapy have tended to become highly focussed and symptom-centred. This has led to the evolution of a fragmented approach to clinical problems, in which symptom-specific treatment has assumed a position of disproportionate importance. There are numerous pitfalls with this approach, not least of which is that, having based the patient's management entirely upon a set of symptoms, failure on the physiotherapist's part to relieve the symptoms may lead to a crisis of confidence and a lack of other treatment options. In this book we argue that specificity is important, but in a *contextual* setting rather than in terms of quasi-precision of technique.

It is hoped that this book will stimulate the reader to evaluate the assumptions underlying their treatments and to consider more critically the issues that are relevant to the clinical decision-making process. The chapters are self-contained and therefore can be read in any order. However, the chapters follow a sequence, leading from a discussion of a philosophy for musculoskeletal physiotherapy, through a review of relevant basic science, to the application of this science in physiotherapy intervention.

The volume makes no pretence to be a comprehensive work covering all issues. The authors have drawn on their own expertise in specific areas and have raised questions that are currently 'key issues'. With developments in physiotherapy practice, it is unlikely that these will be the unresolved key issues in 5 or 10 years time. One can only hope that physiotherapists have addressed some of these problems in that time and from the results of their investigations moved on to generate other questions and problems, which then become the key issues of a subsequent generation.

Jack Crosbie

Acknowledgements

The editors wish to thank the series editors, Janet Carr and Roberta Shepherd, for their help and guidance throughout the preparation of this book. Appreciation is also extended to all the contributors who willingly gave up much time in the preparation of their chapters. Thanks is also extended to the subjects who so kindly agreed to be photographed.

Jack Crosbie
Jenny McConnell

Chapter 1

Optimization and Musculoskeletal Physiotherapy

JACK CROSBIE

INTRODUCTION

If an engineer wishes to solve a problem relating to a complex system involving a number of variable factors, consideration may be given to all the solutions that satisfy the conditions imposed and which allow the equations to balance. After that, some 'solutions' may be rejected because they are clearly fallacious – for example, a guy-rope could not be required to 'push' rather than 'pull'. The engineer's deliberations may produce a single solution, or a number of possible, feasible solutions. The more complex the system, the more likely it is that several solutions may exist.

The human body, with its multiple links and segments, its mono- and poly-articular muscles, its fine balance of motor control and its responsiveness to sensory stimuli, may have many solutions to a problem of movement. Consider the variety of ways in which people perform the same task, depending upon the interaction of factors such as motivation, skill, experience, and so on. Yet many tasks are carried out in remarkably similar ways by healthy people. Walking is a fairly good example of this; certainly everyone tends to display their own characteristic style of walking by which they can be recognized from some distance. However, many features of normal gait are remarkably consistent for normal, healthy populations.

Why should this be so? A key concept in determining the consistency of a functional performance is that of *optimization*, a term suggesting the manipulation of factors to provide the optimum, most efficient combination. Our common understanding of this term has tended to consider that situations in which the physiological cost to the individual has been minimized will be optimal, but this may not always be the case. Occasionally, an activity may be carried out in such a manner that the performer sacrifices physiological economy for the sake of other considerations, such as appearance. Soldiers marching in slow time might be an example of such an activity.

This chapter will consider the idea that physiotherapists treating musculoskeletal conditions need to seek optimal solutions to the clinical problems presented. In this search, recognition of the responses of the body tissues to mechanical and physiological influences and the possibility of altering these responses will be of paramount importance. In addition, the

1

physiotherapist and the patient need to consider the course of action that will produce a solution, and which is practicable and acceptable to both of them. An appreciation of the scientific basis for treatment intervention, and an awareness of those practices not supported by good objective evidence or whose benefits are conjectural, is essential for the development of an optimal treatment plan. The need for such careful, collaborative planning between therapist and patient is the key to a successful outcome. Optimal treatment will recognize the particular constraints effective in each case and will set specific objectives for each patient's management.

BODY MECHANICS AND MOVEMENT CONTROL

Controlled human movement involves a deliberate series of actions, which cause perturbations of body segments with respect to one another. These perturbations initiate segmental motions which, in turn, bring about skilled activity if the sequence and timing of the segmental motions fits a pattern that ensures successful attainment of some predetermined goal. Movement strategies can vary, and the individual often has a number of options available. Lifting a weight can be achieved in a number of ways, with varying demands upon the low back, knees or hips (Figure 1.1). The most appropriate lifting technique may depend upon the individual's previous experience and training. The 'recruitability' of particular muscle groups, the freedom of movement in the relevant joints, the ability to reach the required velocities and so on, may mean that one solution is more suitable than the others, even though some demands may be higher.

In some activities, successful performance may be guaranteed only by

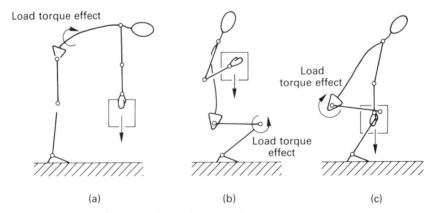

Figure 1.1   Three possible approaches to lifting a weight. In each case the work done is the same (i.e. weight × height lifted) but the demand on the body will be quite different. In (a) the load torque effect will be concentrated in the region of the lumbar spine, in (b) the demand will be greatest at the knees, and in (c) the predominant effect will be at the hips.

employing a very limited variety of movement strategies. The demands of these optimal strategies may require the muscle strength or joint flexibility of specifically trained individuals and, therefore, may make the activity unattainable by the untrained. For example, Olympic gymnasts performing exercise routines on the parallel bars require exceptional ranges of motion in all joints, considerable upper limb and trunk strength and very highly developed motor coordination. The synchronization of the skilled movement sequences required to perform a given manoeuvre is such that the opportunity for variation in the movement strategies is almost zero. Consequently, such gymnasts follow similar training regimens to each other and may have comparable physiques; the execution of the movements will be virtually identical, with little or no *inter-subject variability*. On the other hand, an activity such as dressing does not require a stereotypical set of movements, and so can be performed by people of all ages, body types and, to a great extent, states of health and fitness.

COMPENSATION FOR DISORDERED MOVEMENT

The human body is at the mercy of various forces, gravitational, inertial and muscular, which tend to topple, twist or otherwise move it. An individual naturally tries to balance and integrate these forces in a way that allows a maximum efficiency of effort while achieving the goals set. When the body system experiences some sort of malfunction, or has a new set of external constraints imposed upon it, such as standing up from a different chair height, the body has a new set of variables to deal with and must attempt to optimize the new condition. This often leads to the use of compensatory strategies to allow the performance of the desired function. For example, Saunders *et al.*[1] described biomechanical determinants of human gait, in which the attainment of normal walking was conditional upon the individual's ability to synthesize the six fundamental movements: pelvic rotation, pelvic tilt, knee flexion in the stance phase, normal interactive coordinated movements of the knee and foot, and lateral displacement of the pelvis. They considered that these six fundamental elements interacted, but also stated that the loss of one of these components could be overcome by compensatory activity in the other body regions because of the freedom of movement available.

These compensations are not always 'optimizations', and in fact may create further problems, because the user may habituate to the new condition. Such habituation may lead to the adaptation of contractile and non-contractile tissues, overuse or atrophy of muscle tissue and a resultant imbalance of effective forces, with the development of further musculoskeletal problems.

In some dysfunctional situations compensatory activity may be inevitable. In post-poliomyelitis paralysis, for example, the patient will require to adopt strategies to cope with the disordered movement resulting from factors such as the use of orthoses and reduced muscle activity. These strategies may pose certain problems. If the orthosis, in satisfying the objective of providing mechanical support to an unstable joint, creates a limitation of the available range of movement in adjacent and related joints, overall function may be so disrupted that normal activity becomes impossible or subject to excessive energy expenditure. A high-level paraplegic, with a complete spinal cord lesion, might require such extensive bracing to provide support of the knees, ankles, hips and trunk, and be so encumbered by the construction and weight of the orthoses, that walking, the reason for the brace in the first place, is impossible. In such a situation the solution is as much of a problem as the original impairment.

On the other hand, the lack of stability in a particular joint may have such far-reaching consequences that the optimal solution will involve the sacrifice of mobility. An athlete with instability of the knee caused by ligament disruption, may have to accept that stability of the joint, effected by a brace, can only be achieved at the expense of some loss of range of motion. The alternative is not to continue in the sport, and, perhaps, to walk with a marked limp. The effects of the unstable knee may ramify to adjacent and remote joints leading to increased stresses on body tissues.

In some instances it may not be possible for compensatory strategies to be implemented, because a feature of the action may be so distorted (or absent) that no solution can be found. Subjects with bilateral quadriceps paralysis are unable to rise unaided from a seated position. If they wish to stand up, they must use their upper limbs and, by anchoring their feet and extending their hips while simultaneously pushing down on their hands, 'passively' force the knees into a hyperextended position. Their knees may then be locked in position by the orientation of their weight anterior to the knee joint centre. Such a posture relies on the posterior capsular structures to maintain the knee position. This type of strategy often leads to the assumption of fairly bizarre postures which, in turn, can produce soft tissue adaptation, increased stresses in the connective tissues and the increased risk of associated clinical problems.

If a civil engineer wanted an optimal solution to a construction problem, such as a bridge, several solutions might be found which were perfectly feasible and met all the performance conditions laid down. The ultimate choice might then come down to how easily a solution could be put into place and how cheaply. In the same way, the body systems tend to implement easy and cheap (i.e. non-energetic and non-damaging) solutions. To vary from this is probably to increase the degree of difficulty involved, perhaps requiring a more skilled or more complex response, and usually to increase the energy costs associated with the activity.

People who walk with an abnormal gait have been shown in numerous studies[2,3] to display increased energy expenditure and to walk more slowly and with less continuity of movement. These features suggest that the optimal solution is no longer available to them and that they must be content with a solution which is more difficult, more costly, or both. The consequences of prolonged use of such abnormal movement patterns are extensive, and will be seen in the muscles, the ligaments and the bones themselves.

EFFECTS OF STRESS ON BODY TISSUES

Disturbances within the mechanical system, in the form of local or general changes to body alignment, will produce altered patterns of stress in the body tissues. These alterations, in turn, will elicit physiological and mechanical responses in these tissues, which may not always be in the best interests of the body as a whole. As the local posture changes in response to particular functional demands, so too will the patterns of load transmission in the tissues (Figure 1.2). Increased load transmission produces increased

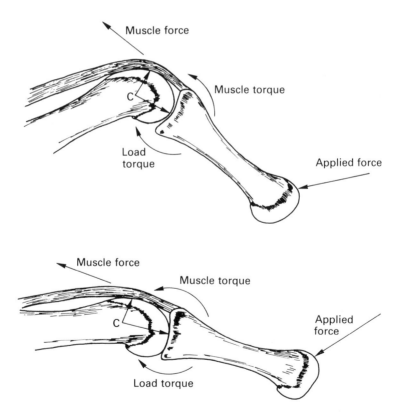

Figure 1.2   Diagrammatic representations of intersegmental joint loading due to the combined effects of applied load and muscular activity. The compressive stress in the underlying bone (C in the figures) will change in both magnitude and location with changing local posture and orientation of the forces.

subchondral bone density, with an increased elastic modulus and a stiffer response to loading. Decreased loading demands will tend to bring about decreased bone density with a reduced elastic modulus. Such changes in the hard tissues are well recognized; for example, Wing and Tredwell[4] have reported increased bone density in the forearms of paraplegics following ambulatory practise with crutches.

As discussed in Chapter 4, the soft connective tissues also demonstrate specific responses to patterns of loading. The more consistent and prolonged the loading characteristic, the more likely it will be to bring about a 'permanent' change in the tissue architecture. Nevertheless, such changes need *not* be permanent, because they are attributable to factors that can vary. The person who, for whatever reason, consistently adopts a local or general posture that is at variance with what is generally defined as 'normal', creates particular patterns of stress in the tissues. These patterns differ from the idealized normal, but they may have become entirely 'normal' for that person and the structure of their tissues may have accommodated to the alteration.

POSTURE

A number of factors may effect such postural changes. Extrinsic influences such as seating design, work-surface height, adequacy of lighting and so on, may over a period impose a new set of postural responses that are at variance with the original posture. The new conditions, however, feel 'normal' to the individual, although the mechanics of their body segmental alignment have undergone marked changes. Intrinsic factors such as growth spurts, self-image or the individual's ability to adapt to changing stresses may also create an internal 'imbalance' of tissue tension, producing a general posture characterized by deviation from the idealized normal.

The extent to which such postural idiosyncrasies require to be modified is a difficult issue. Physiotherapists are encouraged to look at the 'whole person' and to consider how to improve the overall condition of the individual. There has long been a school of thought that stresses the importance of a good postural image and which exhorts the physiotherapist to instil such a characteristic into their patients. The question is, of course, what exactly *is* a good postural image and can it be taught?

The definition of posture tends to be expressed in terms of the relative arrangement of parts of the body,[5] while 'good posture' is "that state of muscular and skeletal balance which protects the supporting structures of the body against injury and progressive deformity irrespective of the attitude in which these structures are working or resting. Under these conditions the muscles will function most efficiently and the optimum positions are afforded for the thoracic and abdominal organs."[6]

The term *posture* is merely a description of the spatial arrangement of the segments with respect to one another. These positions will be determined by internal mechanical factors influencing the adjacent structures, such as the tension in ligamentous and muscular tissues, and external biomechanical demands due to moving or static conditions. These influences will be further modified by the process of habituation. As a postural arrangement becomes established during a particular activity, the individual will tend to consider this as the 'normal' condition. Any departure from this will be registered as 'abnormal'. Anyone who has tried to change the way in which they perform a familiar task will recognize the difficulties inherent in such a change. For example, professional golfers who run into problems with their game often talk about 'rebuilding' their swing. The term is quite appropriate, because the components of the swing need to be learned anew and the sensory perceptions, whether they be classified as proprioceptive or by other terms, need to be felt to be 'normal'. The process builds upon newly learned habits to produce the desired final result, but in the early stages the activity will feel awkward and unfamiliar and they may be tempted to abandon the experiment. The signals from the soft tissues in their arms, trunk and legs are telling their brain that the movement is wrong.

Changes to familiar movement patterns often feel peculiar. Attempts to change habitual patterns of movement, in an attempt to minimize injury risks, are often frustrated by the reluctance of the body to adopt different movement strategies. This problem often thwarts the attempts of physiotherapists working in occupational health to change lifting habits amongst workers. The physiotherapist needs to realize that the relearning process will be slow and may need to be introduced progressively. Mark Twain wrote, "Habit is habit, and not to be flung out of the window by any man, but coaxed downstairs a step at a time".[7]

The key to success will be the extent to which the movement is practised correctly and with feedback within an appropriate setting, and the extent to which the person understands and is prepared to work towards making the change. Such an investment in time and effort has to be seen to be worthwhile; the individual needs to believe that a reward will result from practise.

Changes in local postures in response to musculoskeletal disorders are likely to lead to more general alterations in posture. This relationship may be particularly evident in dynamic situations, where the changing posture of one segment or region has a mechanical effect or consequence in another. The effect of disorder in one part of the body on other regions is well known. Antalgic (pain avoiding) gait, in which the subject's walking pattern is dictated by the desire to avoid weight transmission through one leg because of the pain involved, produces a total body response quite distinct from the local source of the pain.[8]

The body is a multisegmental structure and when one segment becomes immobile, or reduced in its mobility, then adjacent segments may be required to increase their mobility or operate with altered kinematics. Functional activities require the coordinated movement of many body segments and the synergic activity of muscles, some of them quite remote from the site

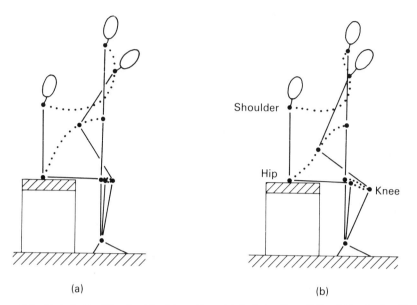

Figure 1.3   Trajectories of the shoulder, hip and knee joints during the activity of rising to stand from sitting in (a) a healthy normal subject and (b) a subject with osteoarthritis of the hip joint. Note particularly the changed pattern of shoulder and knee motion as the subject tries to reduce the torque on the hip by maintaining a more upright posture during the rise. The intermediate position represents the posture at the time of maximum ankle joint dorsiflexion (with thanks to P. Westwood).

of the principal activity. From a diagnostic point of view, the alignment of segments with respect to each other and their trajectories during movement can provide useful information in the identification of abnormality (Figure 1.3). These are manifestations of postural control during a dynamic activity.

STATIC AND DYNAMIC ASSESSMENT

There is a tendency among physiotherapists to neglect objective evidence related to the *dynamics* of human movement and to concentrate instead on the (admittedly more easily acquired) static data, such as the range of joint motion or isometric muscle strength values. This is a regrettable situation, because normal movement requires normal velocity and acceleration characteristics. Joint stiffness is often thought of in terms of the end point of range and a decrease in range of motion. Nevertheless, an increased resistance to motion through range has been described as a feature of certain clinical problems of the musculoskeletal system.[9–11] Impairment of performance may not manifest itself in terms of loss of absolute range of motion, but rather in altered velocity through the range of motion. The investigation of the relationship between displacement and velocity during functional activities such as walking can be very revealing[12,13] (Figure 1.4).

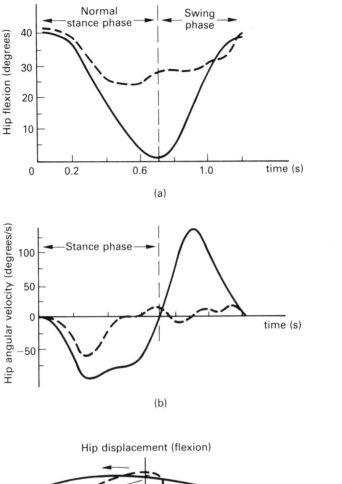

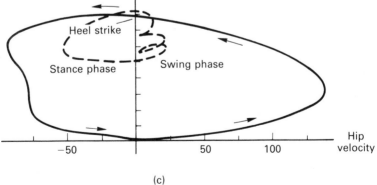

Figure 1.4   (a) Hip joint displacement during one gait cycle. The solid line represents a normal subject, while the broken line represents the pattern produced by a subject with osteoarthritis of the hip joint; (b) hip angular velocity during the same cycle; and (c) phase-plane plot of the hip joint showing the angular velocity of the hip joint as a function of its angular displacement. The smooth nature of the elliptical curve is indicative of a controlled and coordinated activity. The osteoarthritic hip shows marked deviation from the normal pattern.

The theoretical, or idealized normal models proposed and described by investigators in the field of human movement science and used to define desirable, or typical, characteristics of certain actions are extremely valuable to the physiotherapist. It should be realized, however, that the populations used to derive these normative values are not always representative of the entire population, or of the particular sections of the population likely to attend for physiotherapy. In addition, the activities studied in these investigations tend to be simplified in order to minimize the variability of the movements performed. Functional activities are often more complex and subject to more variability, and while the simple model is extremely valuable it must be borne in mind that the analysis presented is limited.

PAIN AVOIDANCE

A common feature in the presentation of disordered movement is the attempt by the individual to avoid pain. The difference between musculoskeletal conditions and problems presenting to physiotherapists working in the areas of neurology or cardiopulmonary disease is the often overwhelming *pain* element. It is this symptom more than anything else that brings the patient to a physiotherapist for treatment.

The pain, or to be more precise the nociception, emanating from a traumatized, inflamed structure, may act as a strong inhibitor to normal movement, producing grossly disordered movement patterns in an attempt to compensate for missing essential components. The effect of such compensatory strategies may lead to adaptation of tissue length, with capsular and ligamentous shortening and loss of sarcomeres in the muscle. Performance of activity becomes depressed because of mechanical impairment due to muscle atrophy or stiff joints, and also by inhibition due to pain or swelling.

Both acute and chronic nociception may manifest in many different ways and may bring about responses from the sufferer which are, on the surface at least, inappropriate considering the actual source of the symptoms. The manifestation of the pain is, nevertheless, quite real to the patient. The ramifications of this most difficult aspect of treatment in musculoskeletal physiotherapy and some bases for physiotherapy intervention are dealt with in detail in Chapter 3.

Immobilization, whether a consequence of deliberate orthopaedic management or as a protective, antalgic mechanism by the patient, has dramatic effects on the body tissues and their mechanical characteristics. Videman[14] has produced strong evidence, reviewing animal experiments, for the theory that immobilization is not only a cause of osteoarthritis, but that it delays the healing process in a traumatized area. Troup and Videman[15] further propose that rest and absence from work following periods of musculoskeletal pain may actually help to prolong the resultant disability and promote degenerative changes in the musculoskeletal system. These concepts are interesting in their own right, but even more so because they suggest that a link exists between biomechanical and psychosocial factors with respect to pain avoidance. These issues are further discussed in Chapter 4.

PREVENTION AND CURE

Any treatment of a problem affecting the musculoskeletal system needs to take into account the remote effects of that problem and to consider the overall needs of the patient. It would seem clear that maintenance of tissue health and prevention of change is more easily attained than recovery after change. Additionally, the context in which the retraining takes place, and the structure of that retraining process, need to be considered. The retraining of specific functional activity of progressive complexity is often part of the rehabilitation programme in neurological problems.[16] It is no less relevant in musculoskeletal rehabilitation, and physiotherapists working in this area might benefit from adopting some of the methods used in the retraining of motor performance. The core of this philosophy is that the patient is a participant and active learner in the process, and that the treatment is directed towards structured practice of functional activities. Such a treatment process does *not* exclude techniques in which the patient is relatively passive and where the physiotherapist applies electrotherapy or manual therapy agents to address particular symptoms. The process, however, is not solely dependent upon those techniques, and the patient is made to feel part of the decision-making and treatment.

The management of a musculoskeletal problem requires consideration of the principles of re-establishing normal tissue compliance characteristics and normal muscle performance. It is apparent that muscle will respond to functionally oriented activity; that, at least, would seem to be the evidence of research to date. On the other hand, the best way of re-establishing normal tissue length and compliance is still unclear. Immobilization does not occur only when plaster splints are applied. The body can effectively immobilize itself by protective muscle 'splinting' and by the adoption of postures designed to minimize motion of painful structures. The rate at which tissue change occurs is surprisingly rapid and is consistent with the period of voluntary immobilization following an acute traumatic episode affecting the soft tissues. It becomes important, therefore, to re-establish tissue length as soon as possible after the period of acute tissue response has subsided.

What is the best way of managing musculoskeletal problems? The diagnostic–treatment model in which labels are sought for particular problems does not provide the physiotherapist with much more information than the labels themselves. To have identified a particular structure as the source of a clinical problem does not, in itself, equip the therapist with a treatment programme. On the other hand, analysis of a motor dysfunction in the context of the limited performance can lead to specific objectives of treatment *and also* give direction to the treatment, encouraging the patient to practise and correct missing or abnormal functional components. This approach is consistent with the principles both of tissue stretching and muscle strengthening. The needs related to connective tissues and muscle are mutually compatible – active use will improve both and increase the chances of recovery.

The decision by the therapist or the patient to forgo compensatory strategies in the pursuit of rehabilitation needs to be made carefully and the advisability of such a course of action thoroughly analysed. The 'optimization strategy' chosen by the patient may have long-term complications, which are not obvious to the patient in the short term. Osteoarthritis of the hip joint will tend to predispose the patient towards a hip flexion deformity which, in turn, leads to an increased lumbar lordosis with associated lumbar spine problems. The deformity also causes a shorter step length, slower gait, decreased exercise tolerance, decreased muscle power, adaptive shortening of soft tissues around the hip and so on. The patient will tend to adopt the 'flexed hip' posture because it is comfortable, decompressing the hip joint and reducing pain, but the physiotherapist should realize the consequences of such a course of action and discuss it with the patient.

The biomechanics of optimal movement during functional tasks should provide the key to musculoskeletal physiotherapy. It is not enough to be familiar with an idealized 'normal', the range of possible, optimal solutions needs to be considered.

The physiotherapist should also understand the concept of essential components of action.[16] These components are the elements that are fundamental in determining the characteristic of the action. Other components may occur, but their purpose is, to an extent, cosmetic. Without the essential components, the action will not proceed in a normal manner. Further to this concept, the physiotherapist needs to have a wide view of the management of the musculoskeletal problem, and not see only the immediate problem without also considering the associated issues.

There is an urgent need for an 'holistic logic' in the management of musculoskeletal conditions. Consider, for example, the situation of a patient recovering from a fracture of the neck of femur. The patient is likely to be elderly and may be quite frail. The priority is to encourage independence, particularly in ambulation, to allow the patient to return home. The patient is probably unable to stand independently from sitting, but may be 'trained' to use a walking frame. There are several flaws in this approach. Firstly, patients cannot practise the walking activity on their own because they cannot achieve the necessary standing position, and consequently the walking training is confined to the times when the physiotherapist is actually present. Secondly, the temporospatial and kinematic characteristics of walking-frame use are so different from 'free normal gait'[17] that such practise as does occur is of doubtful value. Given these limitations, there may be little likelihood of the patient progressing to functional independence.

There is a strong case for the use of functional activity as the cornerstone of management of the musculoskeletal problem which is supported throughout the various chapters in this book. The ability, for example, to rise from sitting to standing and to reverse this motion, may be impeded by any of a number of factors relating to dysfunction of the musculoskeletal system. It may be possible to remedy this problem by encouraging the subject to undertake structured practise of the task. This would certainly be in keeping with the principles of task-specific strength training advocated by Sale *et al.*[18] and others in the case of 'weak' muscles. Prior to this, however, the situation requires to be analysed and action taken to ameliorate any condition or problem which so interferes with the movement that the activity is not likely to be practised correctly.

An arthrodesed ankle joint, in which the talocrural joint is fixed in some plantar flexion, will so interfere with the dorsiflexion needed to bring the body weight forward over the feet, that it may be impossible to rise to the standing position with both feet symmetrically positioned (i.e. side by side). Clearly, in such a situation physiotherapy intervention has no chance of increasing ankle dorsiflexion to allow the motion to proceed. In such a case the only solution would be to allow a compensatory stratagem which permitted the subject to rise. In doing so, however, it should be realized that the economy of the movement may be lost, at least to some extent, and the subject will have a reduced number of options available. Therefore, the patient may have the arthrodesed ankle placed in advance of the sound one. When standing up is complete, the patient may have to make a substantial postural adjustment prior to performing the next action.

Where the ankle is held 'stiff' as a short-term response to pain or trauma, the sitting to standing activity may be used as part of the rehabilitation process. This approach might be appropriate, for example, following rupture of the lateral ligament of the ankle. The characteristic protective posture of the joint limits the patient's dorsiflexion and inversion. With continued protective 'splinting', the patient runs the risk of soft tissue adaptation, thereby resulting in loss of range of motion and weakness of the anterior tibial muscles. The stiff, painful ankle results in a limp during walking, with short step-length on the unaffected side when weight is taken through the painful ankle. Using the functional activity as a focus, with the foot being moved progressively towards the desired position and, indeed, deliberately positioned such that the subject puts more demand on the affected side, the patient will progressively stretch the tight structures under their own control. The activity is able to be practised frequently, and expectations of appropriate frequency and quality of practise can be discussed and agreed to by the physiotherapist and the patient. Similarly, the patient can be encouraged to take a longer step with the intact leg, so stretching the tight structures in the affected ankle and retraining the normal pattern of gait.

Intra-articular cyclic loading has been shown to be of fundamental importance in maintaining the health and nutrition of cartilage.[19] This result is not as likely to occur with passive mobilization, although the intersegmental joint loading patterns of *active motion* will cause compression by the muscles (see Figure 1.2). However, it is not clear what the optimum cycling frequency is, and we do not know the importance of rhythmic patterns of activity in different muscles (as seen in functional activity). Sustained contraction by one muscle group may not be the most desirable as it may increase stress on one region of the articular surface while decompressing others.

Non-contractile tissue relies on the stimulus of normal activity to reorientate its collagen fibres, particularly after trauma has disrupted them. Repeatedly moving a joint through its functional range, while developing normal levels of intra-articular pressure, would be expected to provide such a stimulus to the capsular, intra-capsular and extra-capsular connective tissues and should be beneficial. There is, however, a cautionary note which needs to be mentioned. The presence of mechanical disruption means that a loss of integrity in, or control over, the articulating structures (for example, a fracture in the region) may mean that the pattern of motion in the region is *not* normal. Care needs to be exercised in such situations.

It is surely inappropriate to consider the musculoskeletal system in isolation from the central and peripheral nervous system. Nonetheless, treatment approaches have tended to be concerned only with the identification and labelling of the structure 'responsible' for the symptoms (despite considerable evidence that such precision is invalidated by the complexity of the human response to disease, trauma and associated nociception – see Chapter 3).

Control of motor activity and the performance of function has a great deal to do with an individual's ability to respond to stimuli and to activate optimal conditions for movement. The priority, as far as the physiotherapist is concerned, is to establish an environment in which such optimal conditions can be activated and which will support accurate practise and, hence, accelerate recovery. This task has never been simple, and can now be considered even more difficult, given contemporary awareness of the multivariable nature of musculoskeletal disorders. Despite this, it is essential that the physiotherapist, equipped with an understanding of the operant principles and having scientific bases for treatment, seeks to analyse the situation as accurately as possible, to measure the outcome and to test theories concerning the mechanisms of dysfunction and, consequently, to apply the intervention in a manner consistent with the patient's functional needs and limitations. To do otherwise is to be content with an approach which refuses to acknowledge a growing body of evidence supporting the use of structured, functionally orientated practise in the acquisition and reacquisition of motor skill.

The following chapters in this book consider, in more detail, a number of issues of relevance to musculoskeletal physiotherapy intervention. The remarkable increase in the knowledge base of the movement-related sciences over the past 10 years can be expected to continue. As it does, and as more questions relevant to physiotherapy are answered (and more questions raised) physiotherapists must be willing to revise both their fundamental assumptions and, consequently, their treatment plans. Now may be an appropriate time for physiotherapists to carry out a critical evaluation of interventions in musculoskeletal conditions and to be prepared to adopt radically different philosophies of treatment.

REFERENCES

1. Saunders J.B.D.M., Inman V.T., Eberhardt H.D. (1953). The major determinants in normal and pathological gait. *J. Bone Joint Surg.*, **35A**, 543.
2. Cavanagh P.R., Kram R. (1985). Mechanical and muscular factors affecting the efficiency of human movement. *Med. Sci. Sports and Exerc.*, **17**, 326.
3. Waters R.L., Hislop H.J., Perry J., *et al.* (1978). Energetics: application to the study and management of locomotor disabilities. *Orthop Clin. North Am.*, **9**, 351.
4. Wing P.C., Tredwell S.J. (1983). The weight bearing shoulder. *Paraplegia*, **21**, 107.
5. Galley P.M., Forster A.L. (1987). *Human Movement: An Introductory Text for Physiotherapy Students*, 2nd edn. Melbourne: Churchill Livingstone, p. 86.
6. Kendall H.O., Kendall F.P., Boynton D.A. (1952). *Posture and Pain.* Baltimore: Williams and Wilkins.
7. Twain M. (1894). *Pudd'nhead Wilson: a Tale.* Piccadilly: Chatto and Windus.
8. DuCroquet R., DuCroquet J., DuCroquet P. (1968). *Walking and Limping.* Philadelphia: Lippincott, p. 66.
9. Breger-Lee D., Bell-Krostoski J., Brandsman J.W. (1990). Torque range of motion in the hand clinic. *J. Hand Ther.*, (January–March).
10. Engin A.E. (1985). Passive and active resistive force characteristics in major human joints. In *Biomechanics of Normal and Pathological Human Articulating Joints* (Berme N., Engin A.E., Correia da Silva K.M. eds.). Dordrecht: Martinus Nijhoff, p. 137.
11. Unsworth A., Yung P., Haslock I. (1982). Measurement of stiffness in the metacarpophalangeal joint: arthrograph. *Clin. Phys. Physiol. Meas.*, **3**, 273.
12. Kelso J.A.S., Vatikiotis-Bateson E., Saltzman E.L., *et al.* (1985). A qualitative dynamic analysis of reiterant speech production: phase portraits, kinematics and dynamic modeling. *J. Acoust. Soc. Am.*, **77**, 266.
13. Winstein C.J., Garfinkel A. (1989). Qualitative dynamics of disordered human locomotion: a preliminary investigation. *J. Motor Behav.*, **21**, 373.
14. Videman T. (1987). Experimental models of osteoarthritis: the role of immobilisation. *Clin. Biomech.*, **2**, 223.
15. Troup J.D.G., Videman T. (1989). Inactivity and the aetiopathogenesis of musculoskeletal disorders. *Clin. Biomech.*, **4**, 173.
16. Carr J.H., Shepherd R.B. (1987). *A Motor Relearning Programme for Stroke*, 2nd edn. Oxford: Butterworth-Heinemann Medical Books.
17. Crosbie J. (1991). Biomechanics of walking frame ambulation. *Proc. WCPT Congress, London*, p. 435.
18. Sale D.G., McComas A.J., MacDougall J.D., *et al.* (1982). Neuromuscular adaptation in human thenar muscles following strength training and immobilisation. *J. Appl. Physiol.*, **53**, 419.
19. Bullough P.G. (1981). The geometry of diarthrodial joins, its physiologic maintenance and the possible significance of age related changes in geometry-to-load distribution and the development of osteoarthritis. *Clin. Orthop.*, **156**, 61.

Chapter 2

Physiological Factors Influencing Performance of Skeletal Muscle

MARIA STOKES AND ROBERT COOPER

INTRODUCTION

Complaints common in patients presenting for physiotherapy are skeletal muscle weakness, wasting (atrophy) and fatigue. Some of the mechanisms of these symptoms are poorly understood but obvious causes include injury, immobilization and neuromuscular disease. Weakness may still occur, however, even when a muscle is intrinsically normal, for example, when activation is inhibited by reflex inhibition or by central mechanisms.

The purpose of this chapter is to provide the reader with a fuller understanding of the known causes of muscle weakness, wasting and fatigue, as derived from investigations into their mechanisms. An understanding of the basic neurophysiology of muscle contraction is assumed, and the reader is referred to suitable texts,[1-3] but a brief background section follows, which includes relevant physiological concepts and terminology required to appreciate the specific areas to be discussed.

BACKGROUND PHYSIOLOGY

The chain of electrical, biochemical and mechanical events leading to a voluntary muscular contraction begins with the initiation of impulses in the brain and ultimately results in myofibrillar cross-bridge formation and the production of force[4] (Figure 2.1). Failure or abnormalities of any of the mechanisms in this chain may result in impaired muscle performance.

Muscle energy metabolism

The immediate energy source required during muscle contraction is provided by the splitting of *adenosine triphosphate* (ATP), which can be

16

Brain
↓
Spinal cord
↓
Peripheral nerve
↓
Sarcolemmal membrane
↓
Transverse tubular system
↓
Calcium release
↓
Actin–myosin interaction
↓
ENERGY
↓
Cross-bridge movement
↓
FORCE

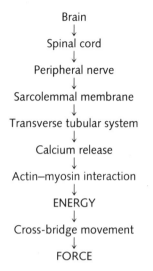

Figure 2.1 Chain of command for muscle contraction. Weakness or fatigue could result from failure at any of the links in the chain. (After Edwards R.H.T. (1983). In *Biochemistry of Exercise* (Knuttgen H.G. ed.). Champain, Illinois: Human Kinetics.)

regenerated under anaerobic and aerobic conditions (Figure 2.2). The energy is used to form myofibrillar cross-bridges and to produce sliding between the contractile filaments, actin and myosin. The substance ATP links energy-yielding and energy-using functions in cells, and details of these reactions have been thoroughly described.[1–3]

The enzyme *creatine kinase* (CK), which catalyses the breakdown of *phosphorylcreatine* (PCr) (Figure 2.2), is of considerable diagnostic value since it leaks out of damaged or abnormal muscle and into the circulation. Therefore, plasma CK is an important marker for muscle damage (see *Minimizing fatigue during treatment*, page 37).

Utilization of ATP

$$\text{ATP} \xrightleftharpoons{\text{Myofibrillar ATPase}} \text{ADP} + \text{inorganic phosphate (Pi)} + \text{ENERGY}$$

Regeneration of ATP
ANAEROBIC

$$\text{Phosphorylcreatine (Pcr)} + \text{ADP} \xrightleftharpoons{\text{Creatine Kinase}} \text{creatine} + \text{ATP}$$

$$\text{Glycogen} + \text{Pi} + \text{ADP} \xrightarrow{\text{Myophosphorylase}} H^+ + \text{lactate} + \text{ATP}$$

OXIDATIVE

$$\underset{\text{(Circulating)}}{\text{Glucose/free fatty acids}} + O_2 + \text{Pi} + \text{ADP} \xrightarrow{\text{Krebs' cycle}} H_2O + CO_2 + \text{ATP}$$

Figure 2.2 Energy pathways involved in the use and regeneration of adenosine triphosphate (ATP) during skeletal muscle metabolism. (⟶ = enzymes or reactions.)

Muscle fibre types

Muscle fibres are usually classified into two main groups, *type I* and *type II*, but alternative names are also used as mentioned below and in Table 2.1.[3] Therefore, fibres are often also named by their metabolic and physiological characteristics, which relate to whether their action is 'fast' or 'slow'.

Type I muscle fibres (slow twitch or ST). These are fatigue-resistant and rely upon a blood-borne supply of glucose, fatty acids and oxygen for their metabolism. They are therefore referred to as *slow oxidative* (SO) fibres and appear red because of the large concentrations of intracellular myoglobin and cytochromes present. Because of their fatigue resistance they predominate in 'postural muscles' such as soleus.

Type II muscle fibres (fast twitch – FT, or glycolytic – FG). These rely on intramuscular energy stores and, although responsible for strength, are highly fatiguable. An example of a fast twitch muscle is the gastrocnemius.

A third type of muscle fibre exists, which forms a type II subgroup and comprises fast but oxidative glycolytic (FOG) fibres known as *type IIa*, whereas the fast glycolytic fibres shown in Table 2.1 are *type IIb*.

All the fibres of a particular motor unit are of the same type, and all human skeletal muscles contain a mixture of types in varying distributions, unlike some animal muscles, which can consist exclusively of one type.[5–7] Further it has been suggested that type I fibres of the human adductor pollicis are faster than those of both the quadriceps femoris and soleus muscles.[6] The result of this is that, despite having very different fibre-typing, the adductor pollicis and quadriceps femoris muscles have very similar contractile properties.[6]

Most human muscles have relatively even proportions of fibre types[5] but highly successful athletes, at the extreme ends of the spectrum, can have a predominance of fibre type in particular muscles. For example, in the quadriceps femoris, type I fibres predominate in long-distance runners but

TABLE 2.1
Classification of muscle fibres

Characteristic	Type I	Type II
Myosin ATPase activity	Low	High
Contraction and relaxation rate	Slow	Fast
Type of contraction	Tonic	Phasic
Muscle function	Stabilizer/Postural	Mobilizer
Fatigue	Resistant	Fast
Myoglobin and capillary content	High–>red	Low–>White
Mitochondria	Many	Few
Metabolism	Aerobic/oxidative	Anaerobic/glycolytic

type II fibres predominate in sprinters.[8–10] Muscle fibre type is thought to be genetically determined, so that training can enhance the contractile properties of a muscle but cannot reverse its intrinsic qualities or fibre type composition.[3] Most studies support this hypothesis but the matter is still controversial (see *Hypertrophy*, page 27).

Muscle strength, power and endurance

The terms strength and power should not be used synonymously. Maximum *strength* refers to the maximal force that can be exerted by a muscle; it is determined by the muscle's cross-sectional area and is related to its ability to produce torque (rotational force) about a joint axis. Muscle *power* refers to the ability to perform work (force or torque and distance moved) per unit of time. In other words, the *rate* of doing work.

Endurance is the ability of a muscle to contract for prolonged periods and is determined by many factors, including the muscle's oxidative capacity, fibre-type content and the intensity of the activity being performed (see *Response to training* and *Minimizing fatigue during treatment*, pages 27 and 37 respectively).

Types of contraction

The terms used to describe muscle contractions during exercise are relatively inaccurate, because they were originally derived from experiments of isolated muscles *in vitro*.[11] During *isometric* (equal length) contractions, resistance is fixed as the tension increases. There is minimal shortening of the muscle and no joint movement occurs. In the intact muscle *in vivo* there is, in fact, some shortening, due to stretching of the series elastic component[1] during tension development, so the contractions are not strictly isometric.

Isotonic (equal tension) contractions involve changes in muscle length with movement of the relevant joint, while the tension remains the same. True isotonic contractions do not usually occur *in vivo* as the load changes, due to gravity and to changes in biomechanics and muscle length, as joint angle alters.[3,11] Contractions involving movement are often referred to as dynamic, as opposed to isometric, contractions.

Isokinetic contractions also involve joint movement with changes in muscle length, but resistance is altered to accommodate the exerted force and specialized equipment is required to perform this type of contraction. The rotational speed is preset on such isokinetic exercise machines, which are known as constant velocity dynamometers (for example the 'Cybex').

Dynamic contractions involve either shortening or lengthening of the muscle as the load is applied. *Concentric* contractions involve muscle shortening so that the two ends of the muscle move towards one another.

Eccentric contractions involve muscle lengthening, when the two ends of the muscle move apart. The physiological consequences of concentric and eccentric exercise differ and are discussed later (see *Voluntary exercise programmes*, page 37). Most functional activities involve combinations of the different types of contraction.

The term *agonist* applies to the muscle group whose contractions are responsible for a given movement. The *antagonist* is the muscle or muscle group whose contractions oppose the movement produced by the agonist.

Recruitment patterns

The force of a contraction depends on the number of active fibres and the force exerted by each fibre. Deployment of motor units during a muscle contraction is termed *recruitment*, which is related directly to the diameter of the neurones. Smaller neurones are easier to recruit and are therefore used first and most often overall. With increasing force of voluntary contractions, progressive recruitment of type I and then type II units occurs.[12] This is the *size principle of motor unit recruitment*, with the small, low-threshold type I units being activated first.

Maximal voluntary contractions (MVC) utilize both fibre types but the involvement of type I and type II fibres varies throughout the contraction. At the onset of MVC all fibres are recruited, initially at high firing rates, but the firing frequency falls as type II fibres drop out due to fatigue.[13,14] It is therefore not possible to maintain maximal force for more than that few seconds. Fibre recruitment is usually determined by the force rather than the speed required for the movement; therefore, strong contractions initially use firing frequencies stimulating mainly the fast type II fibres, whereas weaker contractions use lower frequencies of firing of type I fibres. An exception when recruitment is not determined by the required force is where rapid, unloaded (ballistic) movement is required; in those situations, fast (phasic) motor units are also recruited.[15] The low forces exerted during ballistic movement, despite fast motor unit activity, may be due to counteraction by the antagonists during such movements.[16]

During weak submaximal contractions, force is produced mainly by recruitment rather than by increasing motor unit firing frequency.[17] As force increases, frequency becomes more important and maximal recruitment occurs before maximal force is achieved. This means that any further increase in force after maximal recruitment is due to increased firing frequency.[17]

MUSCLE WEAKNESS AND WASTING

Weakness is defined as inability to produce the required or expected force in rested (i.e. fresh) muscle. It can occur as a result of a defect at specific sites in the chain of command for contraction (see Figure 2.1). Therefore, weakness can result from poor motivation, disorders of the central nervous system, peripheral afferent influences on central pathways, musculoskeletal trauma or disease, peripheral nerve injury, neuromuscular junction failure (such as is found in *myasthenia gravis*) and defects in the muscle itself including inflammatory, toxic, hereditary and endocrine myopathies. It is not within the scope of this chapter to go into all of these mechanisms, so this section is mainly concerned with muscle weakness associated with joint damage,[18] whether due to surgery, trauma or disease, because this is the form of weakness most often encountered by the physiotherapist. The weakness may result from disuse atrophy, which should respond to strength training (see *Response to training*, page 27), or it can occur by reflex inhibition, when the muscle may not fully respond to exercise until the stimulus causing the inhibition is no longer present, which may explain why some patients do not respond to physiotherapy.

Muscle atrophy

Muscles will become wasted as a result of disuse, immobilization, trauma (to the muscle, nerves, soft tissues or skeletal structures), disease and starvation. Atrophy, in the absence of disease, involves shrinkage of muscle fibres with decreases in both contractile and sarcoplasmic proteins. Factors that influence the rate and extent of atrophy include the cause of wasting, muscle length during any period of immobilization and the duration of the immobilization. For example, immobilization with the muscle in a *lengthened* position *increases* its size longitudinally by the addition of new sarcomeres,[19] whereas immobilization in a *shortened* position causes *muscle shortening* due to the loss of sarcomeres.[20] Early mobilization using a mobile cast brace after knee surgery produces less atrophy and more rapid return of function than immobilization in a standard cylinder.[21]

Selective atrophy may occur between agonists and antagonists, different parts of a muscle group and between fibre types. Therefore, joint damage results in greater weakness of its extensors than its flexors, so, in the thigh, quadriceps weakness is greater than hamstring weakness.[22,23] This may be due to reflex activity causing extensor inhibition[24] and flexor facilitation, which occur by a painless reflex.[25–27]

It is commonly believed that selective wasting of the heads of quadriceps occurs, with vastus medialis wasting more than the other three heads. This theory is based on clinical observation and EMG studies, which are conflicting.[28,29] Selective atrophy of parts of the quadriceps might only be apparent as differences in fascial covering make some parts of the thigh more superficial than others, thus making atrophy more visually obvious in parts and also providing less impedance to surface EMG recordings. The few direct data obtained using imaging techniques are also conflicting, with some suggesting a uniform atrophy[30,31] and one study indicating selective atrophy.[32] These discrepancies suggest that the cause of atrophy, including the site and nature of any injury, may determine the type and mechanism of atrophy.

The muscle fibre types do not atrophy at the same rate or to the same extent. Selective atrophy of type II fibres occurs in a number of conditions and is generally thought to reflect inactivity, whereas preferential type I atrophy is uncommon.[33] Some patients with arthrogenous weakness show preferential type II fibre atrophy,[30,34] but type I atrophy is equally common in this situation.[35,36] Upper motor neurone lesions have been shown to produce type II fibre atrophy,[37] and after spinal cord resection the muscle can exhibit exclusively type II fibres.[38] Muscle length and joint mobility may influence selective atrophy.[21] Another possible influence may be selective inhibition of motor neurones,[39] but no conclusive evidence exists to support this suggestion. Knowledge of the type of fibre atrophy may aid selection of appropriate contractions, whether voluntary or stimulated, for rehabilitation exercise.

Muscle size and strength

In any study of muscle wasting or hypertrophy it is essential to use an accurate method for measuring changes in muscle size. Indirect methods of assessment, such as limb circumference using a tape measure, do not enable individual muscles or muscle groups to be studied. Indeed, circumference measurements may seriously underestimate quadriceps wasting[30] and quadriceps growth.[40] This is because the quadriceps waste more than the hamstrings[22] and subcutaneous fat thickness may vary.

Direct measurements of muscle can be made using diagnostic imaging techniques such as computerized tomography (CT),[41] magnetic resonance imaging (MRI)[42] and diagnostic ultrasound scanning.[43,44] Ultrasound is suitable for rehabilitation research as it is cheaper, more accessible than CT or MRI and does not involve exposure to ionizing radiation. There are no known detrimental effects from diagnostic ultrasound, which is the same technique used to examine the human foetus *in utero*. Compound ultrasound B-scanning has been used successfully to measure quadriceps cross-sectional area. The method is very repeatable and has been described in detail elsewhere.[44,45] No formal repeatability studies have been performed for CT or MRI to allow comparison with ultrasound results.[45]

The maximal voluntary isometric force (MVC), which can be generated by the quadriceps, is closely related to its cross-sectional area[46,47] and to body weight.[48] The predictability of this relationship is therefore useful for assessing weakness[18] (Figure 2.3). In patients who do not complain of pain, 'excessive weakness' in relation to muscle size may be due to reflex inhibition[18,49] (see below). Conversely, when pain *is* present, strength testing

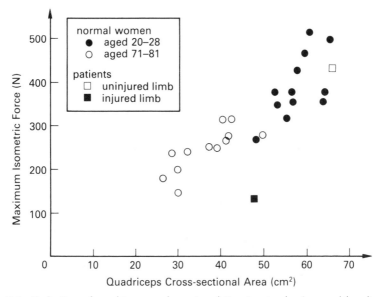

Figure 2.3   Evaluation of quadriceps weakness in relation to atrophy (assessed by ultrasound scanning) in a young female with unilateral knee joint injury. Excessive weakness (■) may indicate reflex inhibition.[18]

may not be appropriate or reliable, so size measurements may be useful for monitoring atrophy or hypertrophy. Some normal young men are stronger than would be expected from the size of the muscle[47,50] but despite this, and other limitations discussed elsewhere,[18] serial measurements of muscle cross-sectional area are still useful for monitoring the effects of rehabilitation or training in this group.

Reflex inhibition due to joint pathology

Reflex inhibition is thought to occur when afferent stimuli from in or around the joint reflexly hamper activation of alpha motor neurones in the anterior horn cell of the spinal cord[49] (Figure 2.4). Inhibition because of pain, or fear of pain, is often confused with painless reflex inhibition. Although pain inhibition involves long loop reflex pathways, there is an obvious voluntary component (although this may not always be conscious) and therefore this phenomenon is not generally referred to as reflex inhibition.

Weakness of the quadriceps femoris is often observed after knee surgery when patients are unable to make maximal contractions no matter how hard they try. This situation, which will be encountered by most physiotherapists, *may* be due to pain inhibition, but pain is not always present.

Reflex inhibition has been studied in animal models for the past

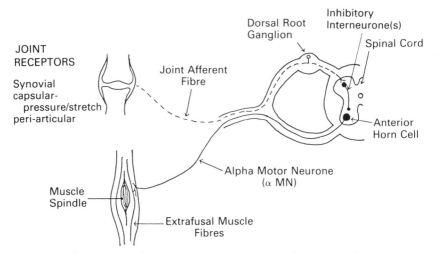

Figure 2.4   Reflex inhibition of muscle activation occurs due to afferent stimuli from in or around a damaged joint. (After Stokes M., Young A. (1987). Proceedings of Xth World Confederation for Physical Therapy, Sydney, p. 416.)

century.[24,25,51] Studies in humans using surface electromyography (EMG) to examine muscle activity have examined the acute effects of controlled knee trauma in two 'experimental' models: a) patients undergoing arthrotomy and meniscectomy;[18] and b) normal subjects with artificially induced knee infusions with saline[52,53]. The reliability and limitations of using EMG to study muscle activity have been discussed by many authors.[54–58] It has been demonstrated that reflex inhibition occurs at a spinal level and is an involuntary mechanism (i.e. not in the subject's control).[49,52] After meniscectomy, inhibition of activation can still be significant two weeks after surgery, despite the fact that patients have been discharged from hospital, are pain free and fully weight bearing.[18] The severity and persistence of the inhibition seen in these patients were somewhat surprising in view of the lack of perceived pain and no obvious effusion, so possible causes were then investigated as follows.

a) Pain. During the first few hours after surgery both pain and inhibition were severe but later some patients still showed severe inhibition despite having no pain. The possibility that later post-operative inhibition is a *learned* response to this early period of pain was examined by infiltration of the knee with a local anaesthetic at the end of the operation. This showed a dose-related effect on pain and inhibition in that a larger dose was needed to block inhibition than to block pain.[59] Since local anaesthesia preferentially blocks the small pain fibres (Aδ and C fibres) before larger fibres carrying sensations of temperature and touch,[60] these results suggest that the inhibitory afferent stimuli were not transmitted along the same nerve fibres as pain impulses. The inhibition does not, therefore, appear to be associated with pain, even when both symptoms are severe.

b) Joint effusion. The presence of fluid in a joint is known to cause inhibition of voluntary contractions,[53] which can be reduced by joint aspiration.[18,49,52] Large effusions are not common after meniscectomy, but when they do occur inhibition is worse and aspiration does not abolish inhibition, so some other factor must also contribute.[18] Infusion of normal knees with saline demonstrated that inhibition can occur due to an involuntary reflex mechanism because reflex activation was also reduced by gradual infusion of the joint.[49,52] This was demonstrated using the quadriceps' H-reflex.[49,52,61,62] Even small volumes of infusion (< 30 ml) caused marked inhibition despite the fact that the fluid was not clinically apparent.[49,52] The inhibition was abolished by removal of the fluid.[49]

c) Intra-articular pressure. In the normal joint, intra-articular pressure alters with joint angle. Highest pressures occur at the extremes of flexion and extension of the knee joint, with the lowest pressure occurring at angles of between 30° and 40°.[63,64] In the presence of joint effusion, much greater pressures are reached.[64,65]

d) Joint position. Inhibition was less in the meniscectomy patients when isometric contractions were performed with the knee in a flexed position[18,66] and this was also previously observed in patients with knee joint effusions.[67] The mechanism of this increased activation is still unclear. It is not thought to be due to changes in intra-articular pressure, because activation increased at all angles of flexion, even those associated with high pressures (as described above). The therapeutic implications of the phenomenon suggest that isometric quadriceps exercises might be more effective at preventing atrophy, increasing strength, or both, if they were to be performed in flexion. If this were shown to be the case, the appropriateness of conventional exercises in extension would have to be questioned.

It is probable that inhibition occurs in all muscles associated with their particular joints. For example, inhibition of biceps brachii has been observed after elbow joint injury.[68] Perhaps inhibition contributes to a variety of conditions, such as the muscle imbalance in idiopathic scoliosis and torticollis in which the cause and effect relationships are unknown. Repeated trauma in sports injury may involve incomplete recovery due to reflex inhibition, causing persistent weakness such as that observed 5 to 10 years after knee surgery.[69] Reduced chest expansion after thoracic and abdominal surgery can cause respiratory complications. The cause of the reduced expansion is not known, but involves reduced activation of the respiratory muscles.[70] Perhaps reflex inhibition of the respiratory muscles results from surgical trauma to these muscles[70] but this requires further investigation. Inhibition may be a necessary 'protective' mechanism after damage to allow the joint to rest but its persistence after the acute phase is counter-productive to controlled therapeutic exercise in the rehabilitation of muscle function.

Effect of ageing on muscle

Muscle performance deteriorates with age, with declines in mass, strength and speed of contraction.[46,47,71–73] The loss of strength has been shown to be proportional to loss of cross-sectional area in the quadriceps muscle.[46,47] The decline begins in the third decade and accelerates after about the age of 50.[74]

The cause of the ageing atrophy is due to actual loss of both muscle fibre types but reduction in fibre size also occurs, mostly of type II fibres.[72,74] Fibres are lost within motor units and there is also a reduction in the number of motor units as indicated by EMG studies[75] and autopsy studies, which demonstrate fewer motor neurones in the spinal cord of older subjects.[76] Whether there is a critical number of motor units below which decline in performance occurs is not known but it is possible that compensatory mechanisms may maintain performance in those individuals who remain active.

The mechanism of the atrophy is thought to be neurogenic, because myopathic changes are rarely seen in normally ageing muscle[77] and neuropathic changes are common in very old individuals.[78] The selective atrophy of type II fibres is thought to be due to inactivity and denervation. The relative contributions of these two factors to atrophy could be examined by comparing life-time athletes with sedentary elderly subjects.

The decline in muscle performance with age appears to occur as a result of intrinsic muscle changes, impaired cardiovascular function and inactivity. The capacity for improvement with training is, however, similar in the young and elderly. This has been demonstrated for both muscle function[79,80] and whole body endurance exercise, which effects increases in maximum oxygen uptake.[81]

The influence of lifestyle and illness obviously play important roles in the decline of muscle with ageing. Disorders such as degenerative joint disease will limit the amount of activity that a person can comfortably perform, and may also cause wasting from reflex inhibition. Other illnesses such as cardiovascular diseases will affect the safety aspect of exercise. These factors will therefore influence the capacity for improvement, which would otherwise be expected to be normal.

Rehabilitation of the elderly after injury can be guided safely by careful assessment and monitoring of muscle function[82] and whole body fitness.[83] These techniques are, of course, useful in sports medicine and in rehabilitation in general. Guidelines for exercise in the older athlete and sedentary person are discussed in a book of review papers.[84] The general opinion is that appropriate exercise is beneficial to the elderly in both health and disease.[85]

Response to training

The response of a muscle to rehabilitation or training depends upon its biochemical and physiological properties, the cause of weakness and the type of exercise used. If weakness is due to disuse atrophy, the muscle should respond to training. Training responses are discussed in Chapter 6, and so are mentioned only briefly here.

Strength. Heavy resistance training preferentially stimulates type II fibre growth, increasing the amounts of contractile proteins, and resulting in improvements in strength. Generally, exercise regimens involve large weights and low repetitions, but strength training is more complex and controversial than this (see Chapter 6).

Endurance. Training for endurance involves low resistance with high repetitions, which stimulates the type I fibres and develops the sarcoplasmic proteins (oxidative enzymes, mitochondrial mass, etc.).

Hypertrophy. Muscle is strengthened by an increase in fibre size, and enhanced motor neurone recruitment and firing rates. These adaptive responses have been reported to differ with age and gender but findings are conflicting. It has been suggested that women of all ages and elderly men increase strength mainly by neural adaptation with some hypertrophy, whereas young men have large increases in muscle size.[79]

The question of whether muscle growth in response to training occurs by an increase in fibre size (hypertrophy) or an increase in fibre number (hyperplasia) is also open to debate. Studies that have demonstrated fibre splitting, suggesting hyperplasia, with intensive resistance training[86,87] have been criticized for their methodology because the splitting may simply reflect fibre regeneration after damage *caused* by the exercise.[3]

Specificity of training. Muscles can be trained using different types of contractions, and the response of the muscle is specific to the exercise used.[88] For example, weight lifting exercises improve the ability to lift weights considerably but do not increase isometric strength to the same extent. This is because the training increases the *skill* factor used during the lifts. Changes in strength are also specific to the length (i.e. joint angle)[89] and speed[90] at which the muscle is trained.

MUSCLE FATIGUE

Fatigue is defined as failure to sustain a given force or power output[91] and may result from impairment at any of the links in the 'chain of command' for muscle contraction[4,92] (see Figure 2.1). This definition could be qualified to accommodate the situation when force is maintained but activation increases to compensate for reduced efficiency.

Classification of fatigue

The causes of fatigue are classified as being central or peripheral in origin[91] (Table 2.2). Central fatigue results from impairment of motor unit recruitment, which may arise from a lack of motivation or altered motor neurone drive. Peripheral fatigue results from failure at or distal to the motor nerve, and is further classified as being high or low frequency fatigue, depending upon the respective stimulation frequency at which force loss occurs. Therefore, high frequency fatigue is the selective loss of force at high stimulation frequencies and involves impairment of either neuromuscular transmission or membrane excitation. Low frequency fatigue is loss of force at low stimulation frequencies and this loss occurs despite normal excitation. Low frequency fatigue is, therefore, due to excitation–contraction coupling failure; a defect *distal* to the sarcolemmal membrane.[93]

The fatiguability of a muscle is increased with muscle atrophy after immobilization or injury. This partly results from the smaller muscle mass having to work harder to perform the equivalent tasks as a normal muscle. However, reductions in type I fibre size[35] and oxidative enzyme concentrations also occur.[21]

Causes of fatigue

a) **Metabolic.** During fatiguing muscular activity, it is extremely difficult to cause ATP levels to fall substantially because of various regenerative processes (see *Muscle energy metabolism*, page 16) and because the onset of fatigue limits the strength of contraction, thereby reducing ATP requirements. This is probably a natural safety mechanism because depletion of

---

TABLE 2.2

Physiological classification of muscle fatigue (based on Edwards.[91])

Fatigue	Characteristics	Mechanisms
Central	Force generated by voluntary effort less than that by electrical stimulation: failure proximal to the muscle	Failure of neural drive: reduction in motor unit recruitment and/or firing frequency
Peripheral	Same force generated by voluntary and stimulated contractions: failure within the muscle	
High frequency (HFF)	Selective force loss at high stimulation frequencies	Impaired neuromuscular transmission and/or propagation of muscle action potential
Low frequency (LFF)	Selective force loss at low stimulation frequencies	Impaired excitation–contraction coupling

ATP would result in a contracture, with muscle damage. However, PCr levels *do* decline and metabolic by-product levels increase, apparently inhibiting further metabolism.[94] Muscle metabolism will not be discussed in detail here, and the reader is referred to relevant literature.[3,94–96]

b) Electrophysiological. The site at which fatigue occurs depends upon the type and intensity of muscular activity; for example, high frequency stimulation will involve failure of neuromuscular transmission. Prolonged activity may result in low frequency fatigue of long duration due to excitation–contraction coupling failure.[93]

High frequency fatigue can cause a tendency for patients to fall, because the muscles are not activated with the high motor unit firing frequencies required for sudden (ballistic) contractions[16] (see *Recruitment patterns*, page 20). This can occur in congenital abnormalities of muscle activation such as *myasthenia gravis*, where high frequency fatigue results from failure of neuromuscular transmission,[91] and *myotonia congenita* and *myotonic dystrophy*, which involve abnormal sarcolemmal membrane excitation.[92,97]

c) Chronic fatigue and pain syndromes. Central fatigue involves failure (either voluntary or involuntary) of neural drive, resulting in a reduced number of functioning motor units and/or firing frequency; for example neurasthenia, hysterical paralysis and conditions where motivation may be impaired such as fatigue syndromes.[91]

The increasing incidence of chronic fatigue and pain syndromes is a considerable problem, both for patients and managers of health-care resources, because treatment is often inadequate, especially due to lack of awareness about these syndromes amongst health professionals. Fatigue syndromes embrace various disorders in which extremely severe subjective weakness and excessive fatigue are the most common symptoms. There is, to date, no convincing evidence of muscle pathology.[98] Other terms include 'effort syndromes',[99] myalgic encephalomyelitis, and 'Royal Free' disease. Symptoms lead to reduced activity and therefore cardiovascular fitness, so that during everyday activities patients may experience symptoms, such as tachycardia and breathlessness, which are usually associated only with intense exercise. These symptoms will inevitably cause anxiety and thereby perpetuate the vicious circle of fatigue and inactivity, creating a descending spiral of disability[100] (Figure 2.5).

The cause of the fatigue in these patients is unknown but studies of peripheral fatigue have shown that the muscle functions normally[101,102] (see *Fatigue during stimulated contractions*, page 31). This suggests that central mechanisms, including psychological factors, can play a role and often complicate fatigue syndromes. An integrated approach to rehabilitation, with a gradual increase in activity together with psychological support, appears to be the one most likely to succeed in these difficult cases.[99]

Chronic musculoskeletal pain, in the absence of arthritis, has been given

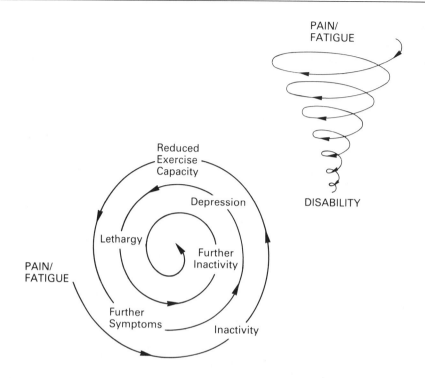

Figure 2.5   Myalgic symptoms and fatigue cause physical inactivity, which can predispose to cardiovascular unfitness and mental lethargy. This leads to exaggerated muscular discomfort on attempted physical activity, which tends to be avoided. Further cardiovascular deterioration is inevitable and the patient enters a 'downward spiral' of immobility, which may end in severe disability and depression. Recovery is directed at gradual but definite increased activity to return affected individuals to a useful lifestyle. (After Cooper R.G. (1991). Rheumatology Now, 7, 18.)

various terms including fibrositis, fibromyositis and fibromyalgia. The symptom profile is almost identical to that of chronic fatigue syndromes and it may be that the two disease entities have common aetiologies.[100] Muscle pain symptoms may also be associated with overuse and/or stress[98,103] and it has been suggested that the term 'regional pain syndrome' be used instead of 'repetitive strain injury'.[103]

d) Respiratory physiotherapy. Studies of respiratory muscle function suggest that fatigue may be an important factor in the development of ventilatory failure.[104–106] When treating patients with chronic respiratory disease it should be remembered that the respiratory muscles may be poorly oxygenated, so over-vigorous or over-long treatment sessions may cause rapid fatigue, which would further reduce the patient's ability for inspiration and coughing. Frequent, short treatment sessions may therefore be more appropriate in these instances. Respiratory physiotherapy is discussed in detail in the companion volume *Key Issues in Cardiopulmonary Physiotherapy* edited by Ellis and Alison.

e) Ageing effects on fatiguability. Muscle endurance declines with age, and this appears to occur to a greater extent in males than females.[107] Following fatiguing exercise, recovery of force of voluntary and stimulated contractions is not impaired.

Changes in muscle energy metabolism have not been conclusively determined. It is generally agreed that anaerobic glycolysis and the activation of certain oxidative enzymes in skeletal muscle are well preserved,[108,109] although, a decline in muscle mitochrondrial respiratory function has been reported.[110]

Capacity for whole body exercise declines, despite the relatively normal muscle metabolism, and the decline occurs before any morphological changes in muscle are evident.[72] The loss of aerobic power is therefore thought to be due to cardiopulmonary changes, notably reduced maximal cardiac output and maximum heart rate.[111] It should be noted, however, that reduced cardiorespiratory fitness is not an irreversible end-point because studies indicate that worthwhile improvements can be obtained by fitness training, which should be fully encouraged.[112]

Consideration of fatigue mechanisms

a) Fatigue during voluntary contractions. In fresh muscle, the same force is generated during maximal voluntary contractions and maximum tetanic stimulation via the motor nerve,[113,114] demonstrating that maximal motor unit recruitment can be achieved by voluntary effort.

Central fatigue. Fatigue obviously occurs during sustained contractions but with maximal effort the best possible force can be maintained reliably with practise, motivation and visual feedback.[92] Central fatigue can be demonstrated during maximal voluntary contractions by superimposing stimulated twitches.[115] If effort is maximal, the stimulation will not further increase force but if effort is submaximal, the twitches do produce more force. This is known as the *twitch interpolation technique*[116,117] and has been used to demonstrate central fatigue in subjects with 'chronic fatigue syndromes'.[101,102]

Peripheral fatigue. The peripheral element of fatigue during voluntary activity is seen when stimulation of the motor nerve does not increase voluntary force.[113] The site of failure appears to be beyond the neuromuscular junction because the muscle action potential of stimulated twitches does not decline during voluntary activity.[118]

b) Fatigue during stimulated contractions. Electrical stimulation of motor nerves enables the central nervous system to be bypassed, allowing peripheral mechanisms to be examined in isolation. Such investigation is relevant to therapeutic electrical stimulation and studies mentioned below illustrate the importance of considering fatigue mechanisms when using stimulation

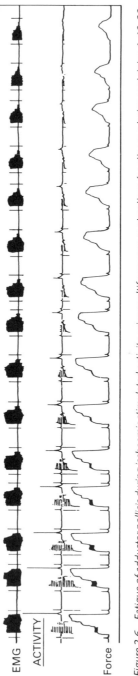

Figure 2.6 Fatigue of adductor pollicis during ischaemic stimulated activity using multifrequency contractions (contiguous trains containing 1, 10, 20, 50 and 100 Hz for 2 seconds each). Contractions were repeated with a 5 second interval between each.[119] Copy of original recordings from one normal subject showing tracings of force, EMG and relaxation rate.

therapy. The adductor pollicis muscle has been studied by many workers because all the fibres may be activated by supramaximal stimulation of the ulnar nerve, and membrane excitation levels can be examined simultaneously using EMG.[48,113,119]

Mechanisms of peripheral fatigue

Fatigue has been studied using various single frequency trains of stimuli and conflicting conclusions have been drawn regarding the mechanisms of the force failure. An investigation[119] used multiple stimulation frequencies in a contiguous train, which produced a characteristic profile in normal muscle[91] and which altered during fatigue[119] (Figure 2.6). This allowed multifrequency examination of fatigue throughout activity and showed that the relationship between change in force and excitation is dependent upon the impulse frequency used.[119] Therefore, high frequencies of stimulation produce EMG changes before any significant fall in force, indicating neuromuscular junction failure, whereas low frequencies cause greater declines in force than excitation, indicating excitation–contraction coupling failure.

Pattern of stimulation frequencies

The rate of fatigue during stimulated activity is influenced by the order in which the frequencies are delivered. With sustained high frequency (e.g. 100 Hz), declines in force occur but when the frequency is reduced (e.g. to 20 Hz), the force increases again.[120,121] High frequency fatigue is thought to be due to increased extracellular potassium (K^+), decreased sodium (Na^+) concentrations, or both, which could be rapidly reversed with a reduction in frequency.[121] When, during stimulated activity protocols, the order of the multifrequency train commences at a high rather than a low frequency within each contraction, low frequency force is greater, and fatiguability reduced, throughout the activity.[122] This effect is due to *post-tetanic potentiation*, which is the enhancement of a low frequency contraction when it is preceded by tetanic (i.e. high frequency) activity.[123–126]

During therapeutic muscle stimulation, impulse frequency must therefore be selected carefully in order to be appropriate for the desired treatment effect. For strength improvements the contraction must be strong enough to stimulate type II fibre growth and hence high frequencies are required, although if the frequency is too high the muscle will fatigue rapidly. For improvements in muscle endurance capacity, low frequency stimulation is required[127] but chronic low frequency stimulation causes normal muscle to become weaker,[128] although strength can be maintained if high frequency bursts are interspersed within the low frequency trains.[128]

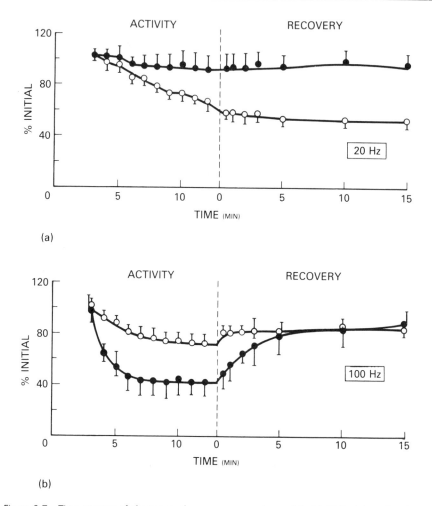

Figure 2.7 Time courses of changes in force, as a percentage of that initially (o), and excitation amplitude, as a percentage of that initially (●), at (a) 20 Hz and (b) 100 Hz during stimulated activity of adductor pollicis.[119] Mean ± 1 sd for nine normal subjects. Activity, without circulatory occlusion, consisted of 50 multifrequency contractions (see Figure 2.6). Measurements were made on every 5th contraction during activity and at various times during recovery. At high frequency (100 Hz) force and excitation have recovered by 10 minutes of rest while at low frequency (20 Hz) fatigue persists despite full recovery of excitation. This phenomenon is termed low frequency fatigue and is thought to represent excitation–contraction coupling failure.[93]

Low frequency fatigue

After prolonged activity, recovery of force may take considerably longer at low than high frequencies.[93] This is demonstrated in Figure 2.7, where there is an obvious separation of force and excitation at 20 Hz but not at 100 Hz.[119]

The possibility of the persistence of the low frequency fatigue effects during subsequent activity was examined by repeating the multifrequency activity protocol illustrated in Figure 2.6 when low frequency fatigue had already been induced.[129] The results showed that fatigue at high frequencies was similar during both series of contractions, whereas fatigue at low frequencies was greater during the second series, despite similar changes in excitation to those in the first series. This confirmation that low frequency fatigue persists during subsequent stimulated activity suggests that the impaired endurance of fatigued muscle during submaximal voluntary activity,[130,131] which involves low firing frequencies,[14] primarily results from peripheral changes at low frequency.[129] Low frequency fatigue is not caused by metabolic factors because energy stores are replenished relatively rapidly following activity.[3,95,96] Therapeutic stimulation at low frequencies will therefore be less effective when the muscle is fatigued, so adequate recovery must be allowed to take place.

Impaired performance of fatigued muscle is associated with increased perception of effort for achieving a given force.[132] Increasing impulse frequency, the number of motor units recruited, or both, will enable the required force to be achieved in the presence of low frequency fatigue. Evidence that such compensation occurs during voluntary contraction is seen with increased total EMG activity for a given force when muscle is fatigued.[133] These alterations in motor neurone drive may be the cause of the increased perception of effort. Low frequency fatigue is more severe and prolonged after eccentric contractions,[131] which have been shown to cause muscle damage[134] (see *Minimizing fatigue during treatment*, page 37).

ENHANCING MUSCLE PERFORMANCE

Knowledge gained from scientific studies can be applied to therapeutic interventions, which aim to optimize muscle performance.

Non-invasive testing

Accurate, objective non-invasive testing of muscle function is essential for research and can be useful for diagnosis and for guiding the clinical management of patients with impaired muscle function. The methods used in the studies discussed in this chapter, however, are not always available to the clinician and strength testing equipment is probably the most accessible to physiotherapists. Detailed discussion of these techniques is contained in contemporary texts.[135]

Compound B-scanning is useful for research but it is too time-consuming and impractical for routine clinical use. Another form of ultrasound, real-time (RT) scanning[136] is cheaper and more accessible than compound scanning. While RT scanning does not enable whole cross-sections of large muscles to be obtained, linear measurements of part of a cross-section can be made. The greater portability of RT scanning, and other advantages, make it suitable to be adopted for routine clinical use and studies are being carried out to prove its validity.

Programmed stimulation of muscle produces a characteristic profile in normal muscle,[48,93,113,119] which can be used to examine fatigue mechanisms and, clinically, to monitor the effects of therapeutic stimulation. Supramaximal stimulation of part of a large muscle can be achieved safely using pad electrodes for such testing.[48,93] Motor nerve stimulation, as would be used for small muscles, would be uncomfortable and potentially dangerous.

It is clear from the stimulation studies discussed that examination of fatigue using force or EMG measurements alone will not identify the true extent or type of fatigue to be elucidated.[119] This is because force and EMG are dissociated when the muscle is fatigued as seen in Figure 2.7[93,96,119,133] and so both parameters need to be measured.

Reduction of inhibitory factors

a) Central. Since pain can cause voluntary inhibition, it should be reduced as much as possible before exercise is attempted. Pain may be severe enough to require medication, or may be reduced by physiotherapeutic techniques such as ice therapy, manual therapy or electrotherapy. Motivation is very important for achieving maximal effort,[92] so encouragement should be provided by the therapist.

b) Peripheral. Reflex inhibition in the presence of joint damage is peripheral in origin, and can be reduced if the stimulus causing the inhibition is removed[18] (see *Reflex inhibition due to joint pathology*, page 23). The potent effect of joint effusion on muscle activity indicates that large, tense effusions should be therapeutically aspirated. The inhibitory reflex mechanism does not appear to operate in the presence of chronic effusions,[137] probably as a result of the capsular tissues becoming stretched and therefore the receptors no longer being stimulated. However, the effect of chronic effusion during functional activity, when large pressure changes occur within the joint,[64,65] needs to be investigated.

Percutaneous muscle stimulation has been shown to prevent muscle atrophy after knee surgery.[138] It may be possible that the effect of reflex inhibition can be bypassed by such a treatment (see *Therapeutic stimulation of muscle*, page 38).

It is not known to what extent the findings from investigations of quadriceps can be extrapolated to other muscles. The observation of biceps inhibition mentioned previously suggests that this might be possible but other muscles and joints must be investigated. Quadriceps is an appropriate experimental model because it is relatively uncomplicated to study and is a very important functional muscle that commonly requires rehabilitation.

Use of feedback to enhance performance

Visual and verbal stimuli[92] can be used to provide feedback to the subject to enhance performance. For example, biofeedback using EMG is commonly used for therapeutic exercise. During strength testing, visual feedback of force output from a monitor can help the subject to achieve maximal force,[92] and submaximal targets can also be maintained more easily. The use of video can also play a role in rehabilitation and sports science to help improve skill in specific manoeuvres.

Specificity of training

The specificity of a muscle's response to training suggests that the main effect of an exercise is on coordination of the muscles involved in that movement (i.e. the neural effect) rather than intrinsic increases in strength.[88] Training for isolated muscles or muscle groups may not therefore be the most appropriate way to improve functional ability, although whether a critical threshold of muscle function should be reached *prior* to functional training is unknown. Since early increases in strength during training are thought to be mainly a result of neural adaptation (see *Hypertrophy*, page 27), perhaps specific training *is* important early on. Rehabilitation programmes should include different types of exercises and specific tasks should be practised to achieve the degree of skill required to carry them out effectively.[88] Specificity of training is discussed in more detail in Chapter 6.

Minimizing fatigue during treatment

Many insights have been gained into minimization of fatigue during treatment as a result of electrophysiological investigations of muscle function.[14,119–124] The principles derived are applicable to both voluntary and stimulated activity.

a) Voluntary exercise programmes. The most obvious situation where fatigue may limit performance is sport, where success depends upon improving fitness (both cardiovascular and muscle endurance) to minimize fatigue. Exercise regimens for improving muscle function have been carefully contrived, and the same principles that apply to sports training also apply to rehabilitation exercise, although the goals and intensity of training will obviously differ. Fatigue should be considered when designing exercise programmes. It is important to select appropriate numbers of repetitions of high or low resistance contractions, depending upon the required effect (i.e. for increasing strength or endurance) and to ensure adequate rest is allowed (see *Low frequency fatigue*, page 34).

Eccentric contractions cause more severe and longer lasting low frequency fatigue than concentric contractions in normal muscle.[131,134] Energy consumption and activation are less but the force exerted per fibre area is greater during eccentric contractions, which are associated with damage and delayed onset muscle soreness.[134] The muscle damage is evident on microscopic examination of biopsy specimens and is detected clinically by a raised plasma CK[134] (see Figure 2.2 and *Muscle energy metabolism*, page 16). While fatigue may be a necessary stimulus to achieve a training effect, 'overtraining' can occur if recovery from fatigue and damage is not allowed,[3] so particular consideration should be given where eccentric exercises are to be used.

It is not clear how *diseased* muscle responds to exercise. The capacity for recovery from eccentric damage may be reduced in diseased muscle. Indeed, it has been hypothesized that eccentric contractions cause progressive muscle damage in proximal myopathies.[139] Care should therefore be taken during exercises to limit the eccentric component until its effects in diseased muscle are known.

b) Therapeutic stimulation of muscle. By the very fact that electrical stimulation is used to study fatigue, it is obvious that therapeutic stimulation will also cause fatigue. The proliferation of the use of muscle stimulation in rehabilitation, despite the lack of scientifically based guidelines for its use, is of major concern.

It has been seen in previous sections that the type and extent of fatigue depends upon impulse frequency and pattern.[119,121,122] The phenomenon of post-tetanic potentiation could usefully be applied to therapeutic stimulation but appropriate frequency patterns need to be established.

Contractions sufficient for some functional movements or to produce increases in muscle strength require large forces, and if these are greater than 30% of the maximal potential force, capillary compression will occur, leading to intramuscular ischaemia.[140] Fatigue is greater under ischaemic conditions;[113,119] therefore brief, intermittent contractions, which allow reperfusion of the muscle, will cause less fatigue than sustained high frequency stimulation or strong voluntary contractions.

Adequate rest periods are required between contractions and treatment sessions because low frequency fatigue impairs subsequent activity[129] but the optimal rest periods for different therapeutic stimulation regimens are not known.

Voluntary contractions are asynchronous, not all motor neurones being recruited at once, whereas stimulated contractions are synchronous. Since only part of a muscle can be stimulated with surface pad electrodes, a smaller population of fibres produce relatively large forces to achieve a given force of contraction. Therapeutic stimulation could be applied more functionally by using multiple electrodes to recruit more fibres and to allow fibres to intermittently rest while the muscle is still contracting, thereby minimizing fatigue.[141]

Beneficial effects of stimulated contractions have been reported in patients with mild neuromuscular disease[142,143] but stimulation of diseased muscle needs careful consideration. Less improvement might be expected in patients with advanced disease in whom penetration of the subcutaneous fat, intramuscular fat, or both, would also pose problems.[144] The question regarding safety of therapeutic stimulation is emphasized by the possibility that certain metabolic myopathies could be present but undiagnosed, such as, for example, myophosphorylase deficiency (McArdle's disease)[145] (Figure 2.2). Stimulation in such patients could cause contracture and muscle damage.[146,147]

Muscle stimulation can be used in paraplegic and tetraplegic patients to enable functional activities to be performed. Muscles must be prepared for receiving functional electrical stimulation (FES) after a period of immobilization. Improvement in muscle endurance capacity can be achieved by low frequency 'conditioning' stimulation.[127] Muscle strengthening regimens then need to be undertaken, particularly if the muscle function involves weight bearing. It is essential to ensure that body weight is not so great as to cause damage during these artificially induced contractions, because force per fibre area is already increased with stimulation.

Electrical stimulation has also been applied to stroke patients,[148] where it was reported to both inhibit unwanted muscle activity and to produce functional contractions. There are many other recorded uses of FES in rehabilitation; these have been reviewed elsewhere.[144,147,149]

Since neural adaptation plays an important role in muscle strengthening, electrical stimulation alone will not regain full function with regard to skill, because it only increases contractile tissues. Electrical stimulation is therefore often used in conjunction with voluntary exercise to initiate contractions.

Before widespread adoption of therapeutic muscle stimulation occurs, whether for normal or diseased muscle, the optimum stimulation parameters need to be determined. These include stimulation intensity (voltage), pulse width, frequency and pattern of frequency, duty cycles and duration of adequate rest periods between treatment sessions. The parameters will obviously depend upon the initial state of the muscle, the type of functional activity in which the muscle is normally involved and the level of performance required, for example to enable an elderly person to regain some independence or an élite athlete to return to competition. Establishment of optimal parameters should help to achieve treatment goals and to prevent overuse or inappropriate use of electrical stimulation.

CONCLUSIONS

This chapter has reviewed the physiological mechanisms of muscle weakness and fatigue, and these should be considered when planning treatment programmes and applying specific techniques.

The cause of weakness will determine the response to treatment so if, for example, a patient does not respond to strength training, the weakness may not simply be due to atrophy; reflex inhibition may be present. Central causes of weakness may be overcome by feedback during exercise as well as verbal encouragement and reassurance.

Muscle fatigue should be considered during treatment and efforts to minimize fatigue (as indicated from studies of fatigue mechanisms) should be made. Central fatigue, such as occurs with chronic fatigue syndromes, can be seriously debilitating and can be improved by gradually increasing activity levels in a supervised programme involving physical and psychological rehabilitation techniques.

The effects of the ageing process on muscle function can be reduced by maintaining an active lifestyle or increased by inactivity and illness. The capacity for improvement in healthy elderly individuals is similar to that in younger subjects, so the elderly should be encouraged to take exercise appropriate to their general state of health.

Accurate, objective techniques for studying the physiological mechanisms of muscle function are essential in rehabilitation research, and some of these have been discussed. Many research techniques can be used to aid clinical diagnosis, and monitoring of muscle, but some of these need to be developed to make them more clinically applicable.

Therapeutic techniques and regimens are often used without guidelines for their most appropriate use, and the effectiveness and safety of some techniques are uncertain. It is essential that findings from basic scientific investigations are used to form a sound basis for clinical studies, and that the findings of these are incorporated into principles underpinning therapeutic practice.

ACKNOWLEDGEMENTS

We wish to acknowledge our collaborators in the studies mentioned, in particular, Professor Archie Young (Royal Free Hospital School of Medicine, University of London) and Professor Richard H.T. Edwards (University of Liverpool). We thank the Department of Social Security, The Muscular Dystrophy Group of Great Britain and Northern Ireland, and ICI Pharmaceuticals Ltd for financial support.

REFERENCES

1. Vander A.J., Sherman J.H., Luciano D.S. (1986). In *Human Physiology: The Mechanisms of Body Function*, 3rd edn. New York: McGraw-Hill.

2. Guyton A.C. (1986). In *Textbook of Medical Physiology*, 7th edn. Philadelphia: WB Saunders.
3. Brooks G.A., Fahey T.D. (1985). In *Exercise Physiology: Human Bioenergetics and Its Applications*, New York: MacMillan.
4. Edwards R.H.T. (1983). Biochemical basis of fatigue in exercise performance: catastrophe theory of muscular fatigue. In *Biochemistry of Exercise* (Knuttgen H.G. ed.). Champaign, Illinois: Human Kinetics, pp. 3–28.
5. Johnson M.A., Polgar J., Weightman D., *et al.* (1973). Data on the distribution of fibre types in thirty-six human muscles: an autopsy study. *J. Neurol. Sci.*, **18**, 111.
6. Round J.M., Jones D.A., Chapman S.J., *et al.* (1984). The anatomy and fibre type composition of the human adductor pollicis in relation to its contractile properties. *J. Neurol Sci.*, **66**, 263.
7. Lexell J., Downham D., Sjöström M. (1984). Distribution of different fibre types in human skeletal muscles. *J. Neurol Sci.*, **65**, 353.
8. Gollnick P.D., Armstrong R.B., Saubert C.W., *et al.* (1972). Enzyme activity and fiber composition in skeletal muscle of untrained and trained men. *J. Appl. Physiol.*, **33**, 312.
9. Gregor R.J., Edgerton V.R., Perrine J.J., *et al.* (1979). Torque–velocity relationships and muscle fiber composition in elite female athletes. *J. Appl. Physiol.*, **47**, 388.
10. Costill D.L., Daniels J., Evans W., *et al.* (1976). Skeletal muscle enzymes and fiber composition in male and female athletes. *J. Appl. Physiol.*, **40**, 149.
11. Gordon A.M., Huxley A.F., Julian F.J. (1966). The variation in isometric tension with sarcomere length in vertebrate muscle fibres. *J. Physiol.*, **184**, 170.
12. Garnett, R.A.F., O'Donovan M.J., Stephens J.A., *et al.* (1978). Motor unit organisation of the human medial gastrocnemius. *J. Physiol.*, **287**, 33.
13. Bigland-Ritchie B., Johansson R., Lippold O.C.J., *et al.* (1983). Changes in motoneurone firing rates during sustained maximal voluntary contractions. *J. Physiol.*, **340**, 335.
14. Marsden C.D., Meadows J.C., Merton P.A. (1983). 'Muscular wisdom' that minimised fatigue during prolonged effort in man: peak rates of motoneurone discharge and slowing of discharge during fatigue. In *Motor Control Mechanisms in Health and Disease* (Desmedt J.E. ed.). New York: Raven Press, pp. 169–211.
15. Desmedt J.E., Godaux E. (1978). Ballistic contractions in fast or slow human muscles: discharge patterns of single motor units. *J. Physiol.*, **285**, 185.
16. Marsden C.D., Obeso J.A., Rothwell J.C. (1983). The function of the antagonist muscle during fast limb movements in man. *J. Physiol.*, **335**, 1.
17. Milner-Brown H.S., Stein R.B., Yemm R. (1973). Changes in firing rate of human motor units during linearly changing voluntary contractions. *J. Physiol.*, **230**, 371.
18. Stokes M., Young A. (1984). The contribution of reflex inhibition to arthrogenous muscle weakness. *Clin. Sci.*, **67**, 7.
19. Williams P.E., Goldspink G. (1971). Longitudinal growth of striated muscle fibres. *J. Cell. Sci.*, **9**, 751.
20. Williams P.E., Goldspink G. (1978). Changes in sarcomere length and physiological properties in immobilized muscle. *J. Anat.*, **127**, 459.
21. Häggmark T., Eriksson E. (1979). Cylinder or mobile cast brace after knee ligament surgery: a clinical analysis of morphological and enzymatic studies of changes in the quadriceps muscles. *Am. J. Sports Med.*, **7**, 48.
22. Arvidsson I., Arvidsson H., Eriksson E., *et al.* (1986). Prevention of quadriceps wasting after immobilisation: an evaluation of the effect of electrical stimulation. *Orthopaedics*, **9**, 1519.
23. Sargeant A.J., Davies C.T.M., Edwards R.H.T., *et al.* (1977). Functional and structural changes after disuse of human muscle. *Clin. Sci. Mol. Med.*, **52**, 337.

24. Charcot J.M. (1889). In *Clinical Lectures on the Diseases of the Nervous System.* Vol. 11. London: The New Sydenham Society, p. 20.
25. Ekhölm J., Eklund G., Skoglund S. (1960). On the reflex effects from the knee joint of the cat. *Acta Physiol. Scand.*, **50**, 167.
26. Sherrington C.S. (1910). On the co-ordination of muscles taking part in the flexion-reflex. In *Selected Writings of Sir Charles Sherrington*, Denny-Brown D. ed. (1979). Oxford: Oxford University Press, p. 163.
27. Lundberg A., Malgren K., Schomburg E.D. (1978). Role of joint afferents in motor control exemplified by effects on reflex pathways from Ib afferents. *J. Physiol.*, **284**, 327.
28. Mariani P.P., Caruso I. (1979). An electromyography investigation of subluxation of the patella. *J. Bone Joint Surg.*, **61B**, 169.
29. Wild J.J., Franklin T.D., Woods G.W. (1982). Patellar pain and quadriceps rehabilitation: an EMG study. *Am. J. Sports Med.*, **10**, 12.
30. Young A., Hughes I., Round J.M., *et al.* (1982). The effect of knee injury on the number of muscle fibres in the human quadriceps femoris. *Clin. Sci.*, **62**, 227.
31. Halkjaer-Kristensen J., Ingemann-Hansen T., Saltin B. (1980). Cross-sectional and fibre size changes in the quadriceps muscle of man with immobilisation and physical training. *Muscle Nerve*, **3**, 275.
32. Gerber C., Hoppeler H., Claassen H., *et al.* (1985). The lower extremity musculature in chronic symptomatic instability of the anterior cruciate ligament. *J. Bone Joint Surg.*, **67A**, 1034.
33. Edwards R., Young A., Wiles C. (1980). Needle biopsy of skeletal muscle in the diagnosis of myopathy and the clinical study of muscle function and repair. *N. Engl. J. Med.*, **302**, 261.
34. Hulten B., Renstrom P., Grimby G. (1981). Glycogen-depletion patterns with isometric and isokinetic exercise in patients after leg injury. *Clin. Sci.*, **61**, 35.
35. Edström L. (1970). Selective atrophy of red muscle fibres in the quadriceps in long standing knee-joint dysfunction: injuries to the anterior cruciate ligament. *J. Neurol. Sci.*, **11**, 551.
36. Young A. (1982). Rehabilitation for wasted muscles. In *Advanced Medicine Vol 18* (Sarner M. ed.). Royal College of Physicians of London. London: Pitman Medical, pp. 138–142.
37. Edström L. (1970). Selective changes in the sizes of red and white muscle fibres in upper motor lesions and parkinsonism. *J. Neurol. Sci.*, **11**, 537.
38. Grimby G., Broberg Z., Krotkiewska I., *et al* (1976). Muscle fibre composition in patients with traumatic cord lesion. *Scand. J. Rehabil. Med.*, **8**, 37.
39. Gydikov A.A. (1976). Pattern of discharge of different types of alpha motoneurones and motor units during voluntary and reflex activities under normal physiological conditions. In *Biomechanics VA* (Komi P.V. ed.). Baltimore: University Park Press, pp. 45–57.
40. Young A., Stokes M., Round J.M., *et al.* (1983) The effect of high-resistance training on the strength and cross-sectional area of the human quadriceps. *Eur. J. Clin. Invest.*, **13**, 411.
41. Häggmark T., Jansson E., Svane B. (1978). Cross-sectional area of the thigh muscle in man measured by computed tomography. *Scand. J. Clin. Lab. Invest.*, **38**, 355.
42. Narici M.V., Roi G.S., Landoni L. (1988). Force of knee extensor and flexor muscles and cross-sectional area determined by nuclear magnetic resonance imaging. *Eur. J. Appl. Physiol.*, **57**, 39.
43. Ikai M., Fukunaga T. (1968). Calculations of muscle strength per unit cross-sectional area of muscle by means of ultrasonic measurement. *Int. Zeitung Physiol.*, **26**, 26.
44. Young A., Hughes I., Russell P., *et al.* (1980). Measurement of quadriceps muscle wasting by ultrasonography. *Rheum. Rehabil.*, **19**, 141.

45. Stokes M., Young A. (1986). Measurement of quadriceps cross-sectional area by ultrasonography: a description of the technique and its applications in physiotherapy. *Physiother. Pract.*, **2**, 31.
46. Young A., Stokes M., Crowe M. (1984). Size and strength of the quadriceps muscles of old and young women. *Eur. J. Clin. Invest.*, **14**, 282.
47. Young A., Stokes M., Crowe M. (1985). The size and strength of the quadriceps muscles of old and young men. *Clin. Physiol.*, **5**, 145.
48. Edwards R.H.T., Young A., Hosking G.P., *et al.* (1977). Human skeletal muscle function: description of tests and normal values. *Clin. Sci. Mol. Med.*, **52**, 283.
49. Stokes M., Young A. (1987). Muscle weakness due to reflex inhibition: future research in different areas of rehabilitation. In *Proceedings of Xth World Confederation for Physical Therapy*, Sydney, p. 416.
50. Maughan R.J., Watson J.S., Weir J. (1983). Strength and cross-sectional area of human skeletal muscle. *J. Physiol.*, **338**, 37.
51. Harding B. (1929). An investigation into the cause of arthritic muscular atrophy. *Lancet*, **i**, 433.
52. Iles J.F., Stokes M., Young A. (1984). Reflex actions of knee joint receptors on quadriceps in man. *J. Physiol.*, **360**, 48P.
53. de Andrade J.R., Grant C., Dixon A. St.J. (1965). Joint distension and reflex muscle inhibition in the knee. *J. Bone Joint Surg.*, **47A**, 313.
54. Desmedt J.E. (1973). *New Developments in Electromyography and Clinical Neurophysiology*. 3rd edn. Basel: Karger.
55. Basmajian J.V., DeLuca C.J. (1985). *Muscles Alive: Their Functions Revealed by Electromyography*. 5th edn. Baltimore: Williams & Wilkins.
56. Ralston H.J. (1961). Uses and limitations of electromyography in the quantitative study of skeletal muscle function. *Am. J. Orthod.*, **47**, 521.
57. Gilmore K.L., Meyers J.E. (1983). Using surface electromyography in physiotherapy research. *Aust. J. Physiother.*, **29**, 3.
58. Yang J.F., Winter D.A. (1983). Electromyography reliability in maximal and submaximal isometric contractions. *Arch. Phys. Med. Rehabil.*, **64**, 417.
59. Shakespeare D.T., Stokes M., Sherman K.P., *et al.* (1985). Reflex inhibition of the quadriceps after meniscectomy: lack of association with pain. *Clin. Physiol.*, **5**, 137.
60. Cousins M.J., Bridenbaugh P.O. (1980). Absorption and disposition of local anaesthetics: pharmacokinetics. In *Neural Blockade*. Philadelphia, Lippincott, p. 54.
61. Hoffmann P. (1918). Uber die Beziehungen der Sehnenreflexe zur willkurlichen Bewegung und zum Tonus. *Zietung Biol.*, **68**, 351.
62. Hugon M. (1973). Methodology of the Hoffmann reflex in man. In *New Developments in Electromyography and Clinical Neurophysiology* (Desmedt J.E. ed.). 3rd edn. Basel: Karger, pp. 227–293.
63. Eyring E.J., Murray W.R. (1964). The effect of joint position on the pressure of intra-articular effusion. *J. Bone Joint Surg.*, **46A**, 1235.
64. Jayson M.I.V., Dixon A. St.J (1970). Intra-articular pressure in rheumatoid arthritis of the knee. III Pressure changes during joint use. *Ann. Rheum. Dis.*, **29**, 401.
65. Levick J.R. (1979). An investigation into the validity of subatmospheric synovial-fluid pressure recordings and their dependence on joint angle. *J. Physiol.*, **289**, 55.
66. Shakespeare D., Stokes M., Sherman K., *et al.* (1983). The effect of knee flexion on quadriceps inhibition after meniscectomy. *Clin. Sci.*, **65**, 64P.
67. Stratford P. (1981). Electromyography of the quadriceps femoris muscles in subjects with normal knees and acutely effused knees. *Phys. Ther.*, **62**, 279.

68. Blockey N.J. (1954). An observation concerning the flexor muscles during recovery of function after dislocation of the elbow. *J. Bone Joint Surg.*, **36A**, 833.
69. Arvidsson I., Eriksson E., Häggmark T., *et al.* (1981). Isokinetic thigh muscle strength after ligament reconstruction in the knee joint. Results from a 5–10 year follow-up after reconstructions of anterior cruciate ligament in the knee joint. *Int. J. Sports Med.*, **2**, 7.
70. Duggan J., Drummond G.B. (1987). Activity of lower intercostal and abdominal muscle after upper abdominal surgery. *Anaesth. Analg.*, **66**, 852.
71. Larsson L., Grimby G., Karlsson J. (1979). Muscle strength and speed of movement in relationship to age and muscle morphology. *J. Appl. Physiol.*, **46**, 451.
72. Grimby G., Saltin B. (1983). The ageing muscle. *Clin. Physiol.*, **3**, 209.
73. Vandervoort A.A., Hayes K.C., Belanger A.Y. (1986). Strength and endurance of skeletal muscle in the elderly. *Physiother. Can.*, **38**, 167.
74. Lexell J., Taylor C.C., Sjostrom M. (1988). What is the cause of the ageing atrophy? *J. Neurol. Sci.*, **84**, 275.
75. Campbell M.J., McComas A.J., Petito F. (1973). Physiological changes in ageing muscles. *J. Neurol. Neurosurg. Psychiatry*, **36**, 174.
76. Tomlinson B.E., Irving D. (1977). The numbers of limb motor neurons in the human lumbosacral cord throughout life. *J. Neurol. Sci.*, **34**, 213.
77. Aniansson A., Grimby G., Hedberg G., *et al.* (1981). Muscle morphology, enzyme activity and muscle strength in elderly men and women. *Clin. Physiol.*, **1**, 73.
78. Tomlinson B.E., Walton J.N., Rebeiz J.J. (1969). The effects of ageing in and cachexia upon skeletal muscle. A histopathological study. *J. Neurol. Sci.*, **9**, 321.
79. Moritani T., De Vries H.W. (1980). Potential for gross muscle hypertrophy in older men. *J. Gerontol.*, **35**, 672.
80. Frontera W.R., Meredith C.N., O'Reilly K.P., *et al.* (1988). Strength conditioning in older men: skeletal muscle hypertrophy and improved function. *J. Appl. Physiol.*, **64**, 1038.
81. Makrides L. (1986). Physical training in young and older healthy subjects. In *Sports Medicine for the Mature Athlete* (Sutton J.R., Brock R.M. eds.). Indianapolis: Benchmark Press, pp. 363–372.
82. Young A., Stokes M. (1986). Non-invasive measurement of muscle in the rehabilitation of masters athletes. In *Sports Medicine for the Mature Athlete* (Sutton J.R., Brock R.M. eds.). Indianapolis: Benchmark Press, pp. 45–55.
83. Wilmore J. H. (1986). Testing the elite masters athlete. In *Sports Medicine for the Mature Athlete* (Sutton J.R., Brock R.M., eds.). Indianapolis: Benchmark Press, pp. 91–107.
84. Sutton J.R., Brock R.M. (1986). *Sports Medicine for the Mature Athlete.* Indianapolis: Benchmark Press.
85. Åstrand P.-O. (1986). Exercise physiology of the mature athlete. In *Sports Medicine for the Mature Athlete* (Sutton J.R., Brock R.M. eds.). Indianapolis: Benchmark Press, pp. 3–13.
86. Goynea W.J. (1980). Role of exercise in inducing increases in skeletal muscle fiber number. *J. Appl. Physiol.*, **48**, 421.
87. Ho K.W., Roy R.R., Tweedle C.D., *et al.* (1980). Skeletal muscle fiber splitting with weight-lifting exercise in rats. *Am. J. Anat.*, **157**, 433.
88. Rutherford O.M. (1988). Muscular coordination and strength training: implications for injury rehabilitation. *Sports Med.*, **5**, 196.
89. Lindh M. (1979). Increase of muscle strength from isometric quadriceps exercises at different knee angles. *Scand. J. Rehabil. Med.*, **11**, 33.

90. Coyle E.F., Feiring D.C., Rotkins T.C., *et al.* (1982). Specificity of power improvements through slow and fast isokinetic training. *J. Appl. Physiol.*, **51**, 1437.
91. Edwards R.H.T. (1981). Human muscle function and fatigue. In *Human Muscle Fatigue: Physiological Mechanisms* (Porter R., Whelan J. eds.). London: Pitman Medical, pp. 1–18.
92. Bigland-Ritchie B. (1981). EMG and fatigue of human voluntary and stimulated contractions. In *Human Muscle Fatigue: Physiological Mechanisms* (Porter R., Whelan J. eds.). London, Pitman Medical, pp. 130–156.
93. Edwards R.H.T., Hill D.K., Jones D.A., *et al.* (1977). Fatigue of long duration in human skeletal muscle after exercise. *J. Physiol.*, **272**, 769.
94. Lewis S.F., Haller R.G. (1986). The pathophysiology of McArdle's disease: clues to regulation in exercise and fatigue. *J. Appl. Physiol.*, **61**, 391.
95. Hultman E., Sjöholm H., Sahlin K., *et al.* (1981). Glycolytic and oxidative energy metabolism and contraction characteristics in intact human muscle. In *Human Muscle Fatigue: Physiological Mechanisms* (Porter R., Whelan J. eds.). Ciba Foundation Symposium No. 82. London: Pitman Medical, pp. 19–40.
96. Miller R.G., Giannini D., Milner-Brown H.S. *et al.* (1987). Effects of fatiguing exercise on high-energy phosphates, force and EMG: evidence for three phases of recovery. *Muscle Nerve*, **10**, 810.
97. Cooper R.G., Stokes M.J., Edwards R.H.T. (1988). Physiological characterisation of the 'warm up' effect of activity in patients with myotonic dystrophy. *J. Neurol. Neurosurg. Psychiatry.* **51**, 1134.
98. Edwards R.H.T. (1986). Muscle fatigue and pain. *Acta Med. Scand.*, **711**(Suppl), 179.
99. Newham D., Edwards R.H.T. (1979). Effort syndrome. *Physiotherapy* **65**, 52.
100. Cooper R.G. (1991). Fibromyalgia, an entity or an excuse? *Rheumatology Now*, **7**, 18.
101. Lloyd A.R., Hales J., Gandevia S.C. (1988). Muscle strength, endurance and recovery in the post-infection fatigue syndrome. *J. Neurol. Neurosurg. Psychiatry*, **51**, 1316.
102. Stokes M.J., Cooper R.G., Edwards R.H.T. (1988). Normal muscle strength and fatiguability in patients with 'effort syndromes'. *Br. Med. J.*, **297**, 1041.
103. Brooks P.M. (1986). Regional pain syndrome: the disease of the 80's. *Bulletin of the Postgraduate Committee of Medicine, The University of Sydney* **42**, 55.
104. Efthimou J., Fleming J., Spiro S. (1987). Sternomastoid muscle function and fatigue in breathless patients with severe respiratory disease. *Am. Rev. Respir. Dis.*, **136**, 1099.
105. Moxham J., Morris A.J.R., Spiro S.G., *et al.* (1981). Contractile properties and fatigue of the diaphragm in man. *Thorax* **36**, 164.
106. Branthwaite M.A. (1989). Mechanical ventilation at home. *Br. Med. J.*, **298**, 1409.
107. Lennmarken C., Bergman T., Larsson J., *et al.* (1985). Skeletal muscle function in man: force, relaxation rate, endurance and contraction time-dependence on sex and age. *Clin. Physiol.*, **5**, 243.
108. Aniansson A., Hedberg M., Henning G.-B., *et al.* (1986). Muscle morphology, enzymatic activity, and muscle strength in elderly men: a follow-up study. *Muscle Nerve*, **9**, 585.
109. Essen-Gustavsson B., Borges O. (1986). Histochemical and metabolic characteristics of human skeletal muscle with respect to age. *Acta Physiol. Scand.*, **126**, 107.
110. Trounce I., Byrne E., Marzuki S. (1989). Decline in skeletal muscle mitochondrial respiratory chain function: possible factor in ageing. *Lancet*, **i**, 637.

111. Mahler D.A., Cunningham L.N., Curfman L.D. (1986). Ageing and exercise performance. *Clin. Geriatr. Med.*, **2**, 433.
112. Smith W.D.F. (1989). Fitness training in the elderly: Canadian experience. *Geriatr. Med.*, **19**, 55.
113. Merton P.A. (1954). Voluntary strength and fatigue. *J. Physiol.*, **123**, 553.
114. Bigland B., Lippold O.C.J. (1954). Motor unit activity in the voluntary contraction of human muscle. *J. Physiol.*, **97**, 17.
115. Bigland-Ritchie B., Jones D.A., Woods J.J. (1979). Excitation frequency and muscle fatigue: electrical responses during human voluntary and stimulated contractions. *Exp. Neurol.*, **64**, 414.
116. Rutherford O.M., Jones D.A., Newham D.J. (1986). Clinical and experimental application of the percutaneous twitch superimposition technique for the study of human muscle activation. *J. Neurol. Neurosurg. Psychiatry*, **49**, 1288.
117. Hales P., Gandevia S.C. (1988). Assessment of maximal voluntary contraction with twitch interpolation: an instrument to measure twitch responses. *J. Neurosci. Methods*, **25**, 97.
118. Bigland-Ritchie B., Kukulka C.G., Lippold O.C.J., *et al.* (1982). The absence of neuromuscular transmission failure in sustained maximal voluntary contractions. *J. Physiol.*, **330**, 265.
119. Cooper R.G., Edwards R.H.T., Gibson H., *et al.* (1988). Human muscle fatigue: frequency dependence of excitation and force generation. *J. Physiol.*, **397**, 585.
120. Davis H., Davis P. (1932). Fatigue in skeletal muscle in relation to the frequency of stimulation. *Am. J. Physiol.*, **101**, 339.
121. Jones D.A., Bigland-Ritchie B., Edwards R.H.T. (1979). Excitation frequency and muscle fatigue: mechanical responses during voluntary and stimulated contractions. *Exp. Neurol.*, **64**, 401.
122. Gibson H., Cooper R.G., Stokes M.J., *et al.* (1988). Mechanisms resisting fatigue in isometrically contracting human skeletal muscle. *Q. J. Exp. Physiol.*, **73**, 903.
123. Ranatunga K.W. (1979). Potentiation of the isometric twitch and mechanism of tension recruitment in mammalian skeletal muscle. *Exp. Neurol.*, **63**, 266.
124. Vandervoort A.A., Quinlan T., McComas A.J. (1983). Twitch potentiation after voluntary contraction. *Exp. Neurol.*, **81**, 141.
125. Blinks R., Rüdel R., Taylor S.R. (1978). Calcium transients in isolated amphibian skeletal muscle fibres: detection with aequorin. *J. Physiol.*, **277**, 291.
126. MacIntosh B.R., Gardiner P.F. (1987). Post-tetanic potentiation and skeletal muscle function: interactions with caffeine. *Can. J. Physiol. Pharmacol.*, **65**, 260.
127. Pette D., Smith M.E., Staudte H.W., *et al.* (1973). Effects of long-term electrical stimulation on some contractile and metabolic characteristics of fast rabbit muscles. *Pflugers Arch.*, **338**, 257.
128. Rutherford O.M., Jones D.A. (1988). Contractile properties and fatiguability of human adductor pollicis and first dorsal interosseus: a comparison of the effects of two chronic stimulation patterns. *J. Neurol. Sci.*, **85**, 319.
129. Stokes M.J., Edwards R.H.T., Cooper R.G. (1989). Effect of low frequency fatigue on human muscle strength and fatiguability during subsequent stimulated activity. *Eur. J. Appl. Physiol.*, **59**, 278.
130. Efthimou J., Belman M.J., Holman R.A., *et al.* (1986). The effect of low frequency fatigue on endurance exercise in the sternomastoid muscle of normal man. *Am. Rev. Respir. Dis.*, **133**, 667.
131. Sargeant A.J., Dolan P. (1987). Human muscle function following prolonged eccentric exercise. *Eur. J. Appl. Physiol.*, **56**, 704.

132. Gandevia S.C., McCloskey D.I. (1978). Interpretation of perceived motor commands by reference to afferent signals. *J. Physiol.*, **283**, 493.
133. Edwards R.G., Lippold O.C.J. (1956). The relation between force and integrated electrical activity in fatigued muscle after exercise. *J. Physiol.*, **132**, 677.
134. Newham D.J., Mills K.R., Quigley B.M., *et al.* (1983). Pain and fatigue after concentric and eccentric muscle contractions. *Clin. Sci.*, **64**, 55.
135. Rothstein J.M. ed. (1987). *Measurement in Physical Therapy.* New York: Churchill Livingstone.
136. Kalama D., Suresh S., Githa K. (1985). Real-time ultrasonography in neuromuscular problems of children. *J. Clin. Ultrasound*, **13**, 465.
137. Jones D.W., Jones D.A., Newham D.J. (1987). Chronic knee effusion and aspiration: effect on quadriceps inhibition. *Br. J. Rheumatol.*, **26**, 370.
138. Eriksson E., Häggmark T. (1979). Comparison of isometric training and electrical stimulation supplementing isometric muscle training in the recovery after major knee ligament surgery. *Am. J. Sports Med.*, **7**, 169.
139. Edwards R.H.T., Newham D.J., Jones D.A., *et al.* (1984). Role of mechanical damage in pathogenesis of proximal myopathy in man. *Lancet*, **i**, 548.
140. Barcroft H., Miller J.L.E. (1939). The blood flow through muscle during sustained contraction. *J. Physiol.*, **97**, 17.
141. Pournezam M., Andrews B.J., Baxendale R.H., *et al.* (1988). Reduction of muscle fatigue in man by cyclical stimulation. *J. Biomed. Eng.*, **10**, 196.
142. Scott O.M., Vrbovà G., Hyde S.A., *et al.* (1986). Effects of electrical stimulation on normal and diseased human muscle. In *Electrical Stimulation in Neuromuscular Disorders* (Nix W.A., Vrbovà G. eds.). Berlin: Springer-Verlag, pp. 125–131.
143. Milner-Brown H.S., Miller R.G. (1988). Muscle strengthening through electric stimulation combined with low resistance weights in patients with neuromuscular disorders. *Arch. Phys. Med. Rehabil.*, **69**, 20.
144. Stokes M.J., Cooper R.G. (1989). Muscle fatigue as a limiting factor in functional electrical stimulation. *Physiother. Pract.*, **5**, 83.
145. McArdle B. (1951). Myopathy due to a defect in muscle glycogen breakdown. *Clin. Sci.*, **10**, 13.
146. Cooper R.G., Stokes M.J., Edwards R.H.T. (1989). Myofibrillar activation failure in McArdle's disease. *J. Neurol. Sci.*, **93**, 1.
147. Cooper R.G., Stokes M.J., Gibson, H., *et al.* (1989). Minimizing fatigue for functional electrical stimulation of muscle. *Clin. Rehabil.*, **3**, 333.
148. Liberson W.T., Holmquest H.J., Scott D., *et al.* (1961). Functional electrotherapy: stimulation of the peroneal nerve synchronized with the swing phase of the gait of hemiplegic patients. *Arch. Phys. Med. Rehabil.*, **42**, 101.
149. Ray C.D. (1978). Electrical stimulation: new methods for therapy and rehabilitation. *Scand. J. Rehabil. Med.*, **10**, 65.

Chapter 3

The Anatomy and Physiology of Nociception

NIKOLAI BOGDUK

INTRODUCTION

Pain is the leading complaint for which patients attend physiotherapists or for which they are referred to physiotherapists. In some cases, a diagnosis may be provided or is otherwise evident, but in many instances, particularly with respect to spinal pain, a pathological diagnosis may be lacking. Physiotherapists are then frequently left to formulate their own diagnosis or to deal with the pain symptomatically. A thorough knowledge of the physiology and pathophysiology of pain is therefore a critical component of the necessary training of a physiotherapist. Moreover, if physiotherapists are obliged by choice or perforce to assume the responsibility for managing a patient's pain, there is no place for having an edited or abbreviated knowledge about the mechanisms of pain. With a proper and comprehensive knowledge, the physiotherapist should be able to: identify those problems amenable to physiotherapy treatment; should understand how these treatments relate to the mechanism of pain involved; and should be able to understand those conditions and mechanisms that really are beyond the province of conventional physiotherapy.

Definition of pain

In defining pain, the Taxonomy Committee of the International Association for the Study of Pain enunciated a critical concept: pain is not a sensation; it is an experience.[1] The Taxonomy Committee defined pain as "an unpleasant sensory and emotional experience associated with actual or potential tissue damage, or described in terms of such damage".[1]

Pain is therefore a psychic or cortical phenomenon. Technically it does not exist unless and until the cerebral cortex receives information that evokes the experience of pain. Under normal circumstances this information arrives along peripheral nerves and the spinal cord from sites of actual or perceived tissue damage. This information, however, is not pain. Although it may lead to the experience of pain, until it does so it is only information about tissue damage. It is information that is evoked by noxious stimuli, and the process by which it is detected and transmitted is referred to as nociception.

Conventional physiotherapy techniques do not deal with pain as such, unless the physiotherapist chooses to become involved with behavioural therapy, whereupon they enter the realm of psychologists and psychiatrists who deal with pain. What physiotherapy does address, at large, is nociception. Techniques are aimed at the disorder causing pain or the transmission of nociceptive information from it. To traditional physiotherapists this distinction may seem only semantic or academic, but international authorities on pain take the distinction seriously and their approach should be heeded by physiotherapists.

Physiotherapy has traditionally evolved in a medical model, which maintains that if a patient's disorder can be cured or if nociception can be blocked, the patient's pain will be relieved. Notwithstanding the faith, conviction and good intentions of physiotherapists operating in this medical model, it is an incomplete picture of pain. Particularly in chronic pain, factors may operate that are beyond simple nociception. These include: the patient's reaction to pain and its symbolic or material meaning to them; reactions of anxiety, anger and depression; the consequences of disability, loss of employment and self esteem; the frustration of not finding a 'cure'; the seemingly perverse yet understandable advantages of a sick-role; the value of 'pain' as a marketable commodity in a medical and legal market place; and the complexities of abnormal illness behaviour as a consequence of pain.[2,3]

These dimensions of the pain problem are beyond the scope of this chapter, but unless physiotherapists are cognizant of them, they risk falling into a trap of seeing only the nociceptive perspective of a patient. By focusing exclusively on the physical, physiotherapists through their own conviction, dedication and energy may unwittingly serve to reinforce the patient's somatosization and disease conviction in situations where a physical approach is actually futile and where emphasis should instead be laid on the patient's coping abilities and responsibility for their own physical, psychological and social problems.

These comments notwithstanding, there is nevertheless a major place for dealing with nociception. Physiotherapy has inherited a reputation of being relatively futile in the management of chronic pain, but this reputation may not be due to the inappropriateness of a physical approach, but rather because nociceptive therapy has been applied poorly, irresponsibly and without insight. Therefore, instead of deferring to contemporary pressures to transfer the treatment of chronic pain to psychologists, there is a place for physiotherapists to improve their performance at a physical level based on a thorough mastery of nociception and its treatment.

NOCICEPTION

The process of nociception involves several components: the detection of tissue damage (referred to as transduction); the transmission of nociceptive information along peripheral nerves; its transmission in the spinal cord; and its modulation. Furthermore it is important to recognize that not all nociception involves peripheral tissue damage. Nociceptive signals can be initiated in damaged or diseased peripheral nerves or in the spinal cord whereupon, although the patient complains of pain in a particular part of the musculoskeletal system, its origin is not in that part but in some component of the peripheral or central nervous system that normally innervates that part.

Transduction

There are only two known mechanisms whereby nociception can be initiated by tissue damage or threatened tissue damage. These are chemical and mechanical mechanisms.

Chemical nociception is a process in which algogenic (pain-producing) chemicals are released in the region of nerve endings capable of detecting noxious stimuli. Chemicals that have this capacity include histamine, serotonin, hydrogen ions, potassium ions, bradykinin and adenosine diphosphate[4] (Figure 3.1). The common feature of these chemicals is that they are typically released by damaged tissue cells or by inflammatory cells. Therefore, chemical nociception occurs only in the presence of actual tissue damage. Related chemicals are the prostaglandins and substance P.

Prostaglandins are synthesized from arachidonic acid, and their synthesis is initiated by phospholipase A, which is an enzyme found in cell membranes and activated by membrane damage. However, generally, prostaglandins are not algogenic. The application of prostaglandins on nociceptive nerve endings does not elicit pain, but prostaglandins have the effect of facilitating the action of other algogenic chemicals on nerve endings.[4]

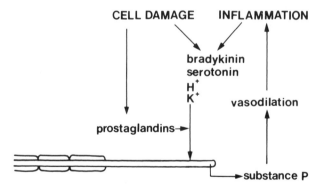

Figure 3.1   Chemical nociceptive transduction. A free nerve ending can be stimulated by chemicals released by damaged cells or inflammatory cells. Prostaglandins facilitate the effect of algogenic chemicals on nerve endings. Substance P, released from nociceptive terminals, promotes vasodilation and reinforces the inflammatory response to tissue damage.

Substance P is involved in nociception as a transmitter substance or co-transmitter released from the central terminals of nociceptive axons (see below). However, it is also released from the peripheral terminals of nerves involved in nociception; i.e. when a nociceptive nerve is activated, substance P is released from both of its ends. In the periphery, substance P does not activate nerve endings, but what it does do is cause vasodilation.[4] In doing so, its role is to promote inflammation and tissue healing. In this respect, it is regarded as part of the *nocifensive system*, which is that component of the reaction to tissue damage designed to promote healing. In this way, nociceptive nerves are not simply transducers of tissue damage but are also actively involved in the repair process.

Mechanical nociception is a less well understood phenomenon because of the technical difficulties involved in studying it. It underlies pain that occurs in the absence of actual tissue damage, but when tissues are being excessively strained. At a microscopic level, mechanical nociception occurs whenever collagen is excessively strained. Therefore, it is the basis for mechanical pain from ligaments, tendons and joint capsules, and from periosteum or skin if and when these are stretched.

The actual mechanism of mechanical nociception is unknown, but would appear to be analogous to the operation of a Golgi tendon organ. In a network of collagen fibres at rest, nerve fibres and nerve endings weave comfortably through the interstices of the network (Figure 3.2a). When the network is stretched, it is deformed and collagen fibres are approximated. This results in nerve fibres and nerve endings being squeezed between them (Figure 3.2b). Presumably, this pressure activates the nerve fibres.

Mechanical nociception and chemical nociception may co-exist, and each of the processes may account for different clinical features. Damaged tissue becomes inflamed and swells. This in turn stretches any surrounding collagen networks. Examples include an abscess whose swelling stretches the dermis and epidermis, or an inflamed joint with an effusion that stretches the joint capsule. The chemicals released by the tissue damage or inflammation result in chemical nociception, which accounts for sustained pain. If the tissue swelling reaches a critical threshold, mechanical nociception may be superimposed. Mechanical nociception is furthermore enhanced in two ways. Firstly, the presence of algogenic chemicals sensitizes nociceptive nerve endings rendering them more easily activated by mechanical stimuli. Secondly, although tissue swelling might of itself cause mechanical nociception by prestressing the collagen network, it renders the nerves in the network more easily stimulated by any additional stimulus. This becomes the mechanism of tenderness.

At rest, normal collagenous tissues have to be deformed greatly before they become painful, e.g. by an arm-lock in wrestling. However, if the tissue is prestressed by swelling, and/or if the nerve endings are sensitized by algogenic chemicals, even the slightest extra, external mechanical stress may activate the nociceptive nerve endings. This is why a tense abscess is painful and exquisitely tender; but once it is lanced, tissue pressure is

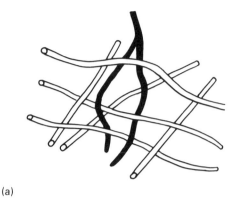

(a)

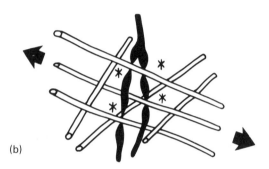

(b)

Figure 3.2   Mechanical nociceptive transduction (a) in a relaxed network of collagen fibres nerve terminals weave comfortably between the fibres, (b) when the network is tensed, the collagen fibres are approximated and squeeze the nerve fibres, which are activated at sites of compression (*).

relieved and the site is markedly less tender. Similarly, swollen joints become less painful and less tender once aspirated.

The above combinations can be viewed as mechanical nociception and chemical nociception in series, i.e. chemical nociception predisposes to mechanical nociception. However, the two processes may also operate in parallel. Tissue damage may occur and attract an inflammatory response. Chemical nociception is established. Meanwhile, uninjured adjacent collagenous tissues remain intact. They are not necessarily swollen or tensed by the inflamed, injured tissues nearby, but their normal function is compromised. This phenomenon typically occurs in partially injured ligaments. The uninjured collagen fibres in the ligament are called upon to continue to bear the load normally borne by the entire, intact ligament. If this load is beyond their physiological capacity, mechanical nociception will ensue. A ligament that has suffered 75% damage has only 25% of its collagen intact. The threshold for mechanical nociception for these remaining fibres is consequently four times less than what it would have been for the entire ligament when it was intact. Under these circumstances, chemical nociception occurs at the injured part while mechanical nociception occurs in parallel at the uninjured, normal part of the ligament.

Nociceptors

In histological terms, there are no specialized receptors involved in either mechanical or chemical nociception. Specialized receptors are involved in detecting touch, temperature and pressure, but the nerve endings involved in nociception are free nerve endings that consist of either a single, tapering terminal or at most, a relatively simple arborization of simple terminals.[5] To date, nociceptors have been classified only according to their response characteristics and the type of nerve fibre from which they are derived.

Nociceptive afferent fibres

The nerve fibres in a peripheral nerve are classified according to their diameter and conduction velocity.[5] Larger fibres conduct more rapidly and as a general rule, amongst myelinated axons the conduction velocity in metres per second is about six times their diameter in microns. When a mixed nerve is experimentally stimulated with an electrode, a compound action potential can be recorded at an arbitrary site some length away from the stimulating electrode. The compound action potential is generated by the sum of the action potentials of the individual axons that constitute the nerve and is referred to as a neurogram. An archetypal neurogram is illustrated in Figure 3.3.

The first action potentials to arrive at the recording electrode are those

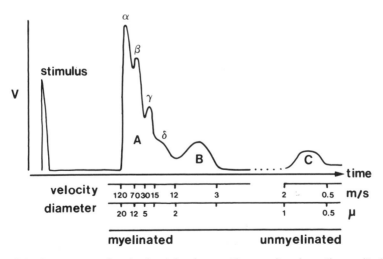

Figure 3.3   A neurogram of a mixed peripheral nerve. The recording shows the amplitude of the compound action potential arriving at a recording electrode after an electrical stimulus that depolarizes the nerve. Three waves of activity are detected (A, B and C) each generated by a respective population of axons characterized by different conduction velocities and fibre diameter. The A wave is subdivided into secondary peaks labelled α, β, γ and δ. The time axis is interrupted because C fibres are very slowly conducting and their action potential arrives much later than the A and B waves.

generated by rapidly conducting axons; the last are those generated by slowly conducting, unmyelinated nerve fibres. The neurogram exhibits three peaks, reflecting three populations of axons each with different conduction velocities. These are known as the A, B and C waves. The C wave arrives very late after the A wave because of the much slower conduction velocity of unmyelinated fibres, and for this reason the time scale at the base of the neurogram has to be interrupted if the A and C waves are to be displayed to scale on the same diagram (Figure 3.3). The A wave is marked by four secondary peaks generated by subclasses of A fibres known as Aα, Aβ, Aγ and Aδ fibres, each with a slightly slower conduction velocity.

B fibres are preganglionic, sympathetic neurones. Aα and Aγ fibres are efferent, motor fibres to skeletal muscle. Sensory fibres create the Aβ, Aδ and C waves.

Aβ fibres are typically sensory axons conveying sensations of touch, vibration, pressure and proprioception. They are not involved in nociception. Nociceptive axons are found only in the Aδ and C class of fibres, although these classes are not exclusively nociceptive.

The Aδ class of fibres includes axons that are not nociceptive. These are axons that innervate low threshold mechanoreceptors (receptors that respond to mechanical stimuli such as light pressure or touch) and thermoreceptors, which respond to temperatures that are not perceived as noxious (Figure 3.4). These neurones exhibit a graded response to non-noxious stimuli, i.e. as the strength of the stimulation increases, their frequency of discharge increases, but at certain thresholds short of what would be noxious, they either cease to be activated or they do not increase their frequency of discharge further (Figure 3.4). Consequently, they cannot code for stimuli in the noxious range.

Nociceptive neurones of the Aδ class differ from these aforementioned fibres with respect to their threshold of stimulation and their response characteristics. They innervate receptors referred to as high-threshold mechanoreceptors or high-threshold thermoreceptors, meaning that they are activated by strong mechanical or thermal stimuli. What renders them nociceptive is that they continue to increase their frequency of discharge as the stimulus becomes noxious, i.e. they exhibit a graded response into the noxious range (Figure 3.4). Nociceptive neurones therefore operate in both the noxious and non-noxious range. No particular type of fibre is exclusively nociceptive. Furthermore, a variety of fibre types is capable of being nociceptive, depending on the type of stimulus to which it is sensitive and its threshold (Figure 3.4). Therefore, there is no such thing as a 'pain fibre' in the peripheral nervous system. All nociceptive neurones in the Aδ class subserve some other sensory function, such as pressure, touch or temperature, in addition to being nociceptive.

In experimental animals, C fibres occur with different response characteristics. However, in humans all C fibres have been found to be polymodal, i.e. they respond to a variety of stimuli including mechanical, thermal and chemical stimuli. These fibres are activated by innocuous and noxious

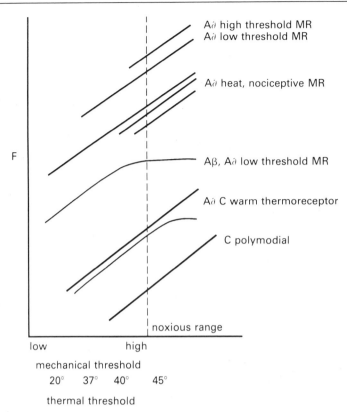

Figure 3.4 *Response characteristics of selected sensory fibres. The graph shows the thresholds and relative frequency of discharge (F) of nerve fibres connected to different types of receptors. The abscissa is calibrated for both mechanical thresholds and thermal thresholds. Low threshold mechanoreceptors (MR) are activated by light mechanical stimuli. High threshold receptors are activated by strong stimuli. Nociceptive fibres are those that exhibit a graded frequency response in the noxious range regardless of the nature of the stimulus or the threshold of the receptor. Non-nociceptive axons do not exhibit a graded response in the noxious range. (Based on* Price D.D., Dubner R. (1977). Pain, **3**, 307.)

stimuli in a graded fashion and are therefore nociceptive. C fibres are the smallest of the axons in peripheral nerves but they are the most numerous. Therefore, C fibres are the predominant type of nociceptive neurone.

Central connections

The majority of sensory nerves enter the central nervous system through a dorsal root of a spinal nerve. However, some enter through the ventral root, their impulses travelling in the opposite direction to those of efferent fibres that constitute the major proportion of this root. Ventral root afferents have only started to be studied and the details of their central connections remain largely unexplored, but the connections of conventional dorsal root afferents are well determined.[5]

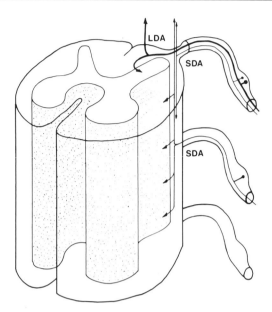

Figure 3.5   The projections of primary afferent fibres in the spinal cord. Large diameter afferent fibres (LDA) are located dorsally (i.e. medially) in the dorsal root as it approaches the spinal cord. They sweep around the medial aspect of the dorsal horn to pass up the posterior columns but collaterals enter the base of the dorsal horn. Small diameter afferents (SDA) are located laterally in the dorsal root. Each divides into ascending and descending branches that travel in the dorsolateral tract and which send collateral branches into the grey matter of the dorsal horns 1–3 segments above and below the segment of entry of the primary afferent.

Within a dorsal root, as it approaches the spinal cord, small diameter afferent fibres (Aδ and C fibres) segregate from large diameter afferents (Aβ fibres). Small diameter afferents assume a more lateral position in the terminal portion of the dorsal root leaving the large diameter afferents more posteriorly (Figure 3.5). The large diameter afferents sweep around the posterior aspect of the dorsal horn continuing mainly up the posterior columns but also sending collateral branches into the middle of the dorsal horn at the levels of laminae IV, V and VI of the grey matter of the dorsal horn.

Small diameter afferents assume a more elaborate course. First, each fibre divides into ascending and descending branches that respectively pass rostrally and caudally along the outer surface of the apex of the dorsal horn in what is known as the dorsolateral tract (of Lissauer)[5] (Figure 3.5). Each branch ascends or descends for one to three segments. Along the entire intersegmental course of each branch, multiple collateral branches pass into the grey matter of the adjacent dorsal horn (Figure 3.5). As a result of this extensive ramification, a given small diameter afferent fibre relays not just to one but to multiple segments at, above and below the segment at which it enters the spinal cord.

Within the spinal cord, second-order neurones capable of transmitting nociceptive information to higher centres are located at two main sites: in

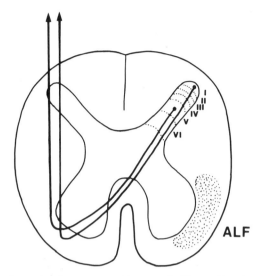

Figure 3.6   Second-order nociceptive neurones. Neurones that transmit nociceptive information in the spinal cord have cell bodies located in lamina I and lamina V of the dorsal horn. Their axons cross the midline in the anterior white commissure and ascend through the spinal cord in the anterolateral funiculus (ALF).

lamina I of the dorsal horn and in lamina V (Figure 3.6).[5] Other nociceptive neurones are located in lamina VII but have not been studied in any detail.

The nociceptive neurones of lamina I are large neurones. They are called marginal neurones because they occupy the marginal lamina of the dorsal horn and lie close to the junction of lamina I and lamina II. These neurones form axons that pass up the anterolateral funiculus of the spinal cord to the thalamus and reticular formation, as do the axons of nociceptive neurones of lamina V (Figure 3.6 and 3.7).

Lamina V neurones are not exclusively nociceptive. They respond to a variety of stimuli including innocuous, mechanical stimuli as well as noxious stimuli. For this reason they are called wide dynamic range (WDR) neurones. The marginal neurones of lamina I are relatively more specific. Most are nociceptive-specific, responding only to noxious stimuli.

Connections are made between small diameter afferents and the neurones in lamina V and lamina I in a variety of direct and indirect ways (Figure 3.8). Direct connections are made on the dendrites of lamina I neurones by small diameter afferents largely of the Aδ class that respond to mechanical stimuli. C fibres and other Aδ fibres terminate on small interneurones in lamina II.

Lamina II neurones consist mainly of two types: islet cells and stalked cells. Stalked cells are excitatory interneurones that receive the terminals of small diameter afferents either directly in the form of conventional axo-dendritic synapses, or in a more complicated fashion in complex structures known as glomeruli (Figure 3.8).

The axons of stalked cells are distributed within their segment of origin

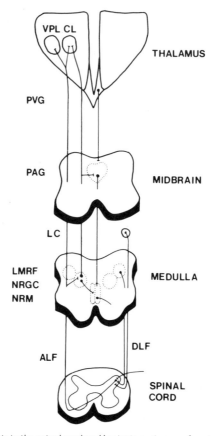

Figure 3.7   Nociceptive tracts in the spinal cord and brainstem. Axons of nociceptive neurones in the dorsal horn pass upwards in the anterolateral funiculus to reach the ventral posterior lateral nucleus (VPL) and centralis lateralis nucleus (CL) of the thalamus. Collaterals from the ascending axons pass to the nucleus reticularis gigantocellularis (NRGC) in the medulla which in turn also projects to the CL nucleus. Descending pathways reach the dorsal horn along the dorsolateral funiculus (DLF) and arise in the nucleus raphe magnus (NRM), the lateral medullary reticular formation (LMRF) and the locus coeruleus (LC). Their activity is controlled by the NRGC, the peri-aqueductal grey matter (PAG) and periventricular grey matter (PVG).

and between segments. Some axons leave the dorsal horn and ascend or descend in the dorsolateral tract to enter lamina I of adjacent segments. Either way, the axons typically terminate on the dendrites of a marginal neurone, thereby completing the connection to a second-order nociceptive neurone.

Other stalked cells relay to the dendrites of WDR neurones of lamina V, which spread from lamina V into the outer reaches of lamina III (Figure 3.8). The same WDR neurones also receive terminals from large diameter afferents sweeping in from the posterior columns. Consequently, the WDR neurones receive both nociceptive and non-nociceptive information, and the neurone is capable of transmitting either type of information. The difference in information is coded by frequency; nociceptive information typically being transmitted as a high frequency discharge, whereas non-nociceptive information has a lower frequency. Therefore, higher centres receiving the axon from a WDR neurone can distinguish the nature of the information being relayed to them on the basis of frequency.

WDR neurones are connected so that they preferentially transmit non-nociceptive information. The terminals of a large diameter afferent aiming for a WDR neurone send collaterals to interneurones that inhibit those dendrites of the WDR neurone that receive axons from nociceptive stalked cells (Figure 3.8). By inhibiting those dendrites the large diameter afferent fibre prevents the WDR neurone from receiving nociceptive information while still allowing the WDR neurone to respond to non-nociceptive information along other dendrites. This circuit provides a means for non-noxious stimuli to inhibit the transmission of noxious stimuli.

Islet cells in lamina II and other interneurones in lamina I have inhibitory functions related largely to descending modulatory pathways (see below). Lamina I interneurones exert postsynaptic inhibition of marginal neurones

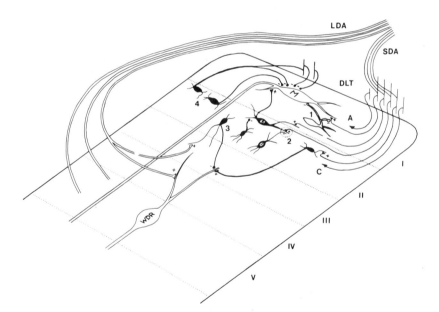

Figure 3.8   Nociceptive connections in the dorsal horn. Large diameter afferents sweep around the medial aspect of the dorsal horn and enter laminae IV, V and VI. They terminate on dendrites of wide dynamic range neurones (WDR) located in lamina V. After travelling along the dorsolateral tract, small diameter afferents enter the dorsal horn. Aδ fibres terminate on dendrites of marginal neurones located in lamina I. Other Aδ fibres and C fibres terminate on dendrites of excitatory interneurones, stalked cells (s), in lamina II. The axons of stalked cells relay to marginal neurones and to WDR neurones. Inhibitory neurones in lamina I (1) provide postsynaptic inhibition of marginal neurones and presynaptic inhibition of primary afferents. Inhibitory neurones in lamina II, islet cells (i), control the interaction between primary afferents and stalked cells in complex structures known as glomeruli (2). Other inhibitory interneurones (3) are activated by large diameter afferents and inhibit the dendrites of WDR neurones that receive nociceptive information. Further interneurones (4) exert inhibitory effects on the cell bodies of marginal neurones both within the same segment and at adjacent segments.

and presynaptic inhibition primary nociceptive afferents (Figure 3.8). Inhibitory interneurones of lamina II send axons within their segment that synapse with the cell bodies of marginal neurones. Other branches of these axons enter the dorsolateral tract to reach marginal neurones at adjacent segmental levels.

These seemingly complex connections have an important role to play in sensory discrimination but they also underlie the mechanism of pain relief of certain therapeutic techniques such as acupuncture, transcutaneous electrical nerve stimulation and vibration (see below).

Transmitter substances

Within the central nervous system, small diameter afferent fibres release a variety of transmitter substances.[4] It is not clear whether nociceptive axons release a single or a variety of transmitter substances, or what this substance is. However, it is known that they release glutamate and a variety of neuropeptides, which are chemicals consisting of chains of amino acids of different lengths and composition. Some authorities believe these neuropeptides to be the transmitter substances of nociceptive axons, others believe they are only co-transmitters and the principal transmitter substance has yet to be discovered. The reluctance to acknowledge neuropeptides as the principle transmitter substances of nociceptive axons is based on the fact that when released onto second-order neurones these substances have a slow onset of action, insufficiently fast for them to account for the speed of detection of noxious stimuli. Nevertheless, neuropeptides have profound affects on nociceptive neurones that suggest that they are involved in the nociceptive process in some way.

The most studied neuropeptide is substance P, a chain of 11 amino acids. This substance is found in many primary afferent fibres and when released on to the dendrites of second-order nociceptive neurones it excites them strongly. Other neuropeptides putatively involved in nociception include cholecystokinin, somatostatin, calcitonin, gene-related peptide, vasoactive intestinal polypeptide and neuropeptide Y; there is some evidence to suggest that nociceptive afferent fibres from different locations and responding to different stimuli contain different peptides. Therefore, chemoreceptive afferent fibres from viscera will have a different neuropeptide from that found in mechanoreceptive nociceptors from skin. This proposition is still being explored.

Ascending pathways

Traditional teaching has maintained that 'pain' travels in the spinal cord along the lateral spinothalamic tract whereas the anterior spinothalamic tract conveys only pressure and touch sensations. In the face of contemporary knowledge of physiology this distinction is no longer appropriate. The axons of lamina I and lamina V neurones are scattered throughout the regions occupied by what formally would be known as the lateral and anterior spinothalamic tracts. They occupy most of the anterolateral quadrant of the spinal cord and constitute what is referred to as the anterolateral funiculus[5] (Figure 3.6). Within this funiculus the axons of WDR neurones convey nociceptive information as well as non-nociceptive information. There is no segregation of the two types of information. The difference is dictated only by the frequency of discharge in the axon, and not by the type or location of the axon.

Most axons of lamina I and lamina V cross the midline in the anterior white commissure before passing rostrally along the spinal cord, whereupon nociceptive information passes along the spinal cord on the side opposite to that on which it entered the central nervous system. However, some axons do pass ipsilaterally, so that nociceptive transmission is not exclusively contralateral. Other pathways for nociceptive information exist in the spinal cord located in the depth of the lateral funiculus and in the posterior columns, but the details of these pathways have not been explored; additionally, their clinical significance has not been determined beyond noting that pathways other than the anterolateral funiculus exist.[5]

The anterolateral funiculus contains axons destined to reach the brainstem reticular formation and others directed to the thalamus. The former are referred to as spino-reticular fibres and the latter, spino-thalamic fibres. In addition to reaching the thalamus, spino-thalamic fibres send collaterals into the reticular formation and these collaterals may also be referred to as spino-reticular (see Figure 3.7).

Within the thalamus, spino-thalamic fibres have two principal destinations: the ventral posterior lateral (VPL) nucleus and the nucleus centralis lateralis. The VPL nucleus is not particularly involved with the aversive, noxious aspects of the information received. This nucleus projects to the parietal lobe of the cerebral cortex, which is responsible for determining the location of the origin of the stimulus. The aversive and emotive nature of pain is triggered by the centralis lateralis nucleus whose connections are not well known, but are presumably to the limbic system of the brain, which is that portion responsible for pleasurable and aversive sensations and their associated visceral and emotional features. In other words, the projections of the centralis lateralis nucleus are responsible for the suffering aspects of pain.

Spino-reticular fibres have diverse connections within the brainstem. Some of these involve activation of descending pathways (see below). Others involve activation of brainstem centres that control respiratory and cardiovascular function, and are responsible for the cardiovascular reactions to nociception. Still others constitute upward, polysynaptic relays to the centralis lateralis nucleus that reinforce the effects exerted on this nucleus by spino-thalamic tracts.

Some authorities have in the past maintained that the anterolateral funiculus may be divided into an outer, lateral spino-thalamic tract or neospino-thalamic pathway and a deeper, medial spino-thalamic tract or paleospino-thalamic pathway.[7] The principal distinction is that the neospino-thalamic pathway predominantly has direct connections to the thalamus whereas the paleospino-thalamic pathway maintains polysynaptic connections via the reticular formation. This distinction is largely conceptual because it does not correlate with any identifiable segregation of nociceptive axons on anatomical grounds. Rather, the distinction emphasizes the two routes by which axons in the anterolateral funiculus may reach the thalamus – a direct one and an indirect one via the reticular formation.

The reason for maintaining this distinction is that it is believed that the neospino-thalamic pathway receives information predominantly from skin and from Aδ fibres, whereas the paleospino-thalamic pathway receives information from deep structures and predominantly (although not exclusively) from C fibres. The paleospino-thalamic pathway is the phylogenetically older pathway (hence its name). It subserves nociception from internal structures. The neospino-thalamic pathway is more recently evolved and subserves nociception from those parts of the body that encounter the external environment.

External noxious stimuli delivered to skin pose an avoidable threat to the body. It therefore makes sense that information about such stimuli would be delivered directly to the thalamus through the neospino-thalamic pathway. Furthermore, this pathway is highly organized somatotopically, i.e. there is faithful coding within the entire system for the origin of the stimulus. Consequently, the information is well localized and reaches the brain rapidly, whereupon it can be analysed and any external threat avoided.

In contrast, the paleospino-thalamic pathway is poorly organized somatotopically. It involves multiple synapses as a result of which the origin of the stimulus is lost, although its aversive nature is nevertheless preserved. There is no need for accurate localization within this system, because there is little benefit to be gained in creating and maintaining a complex system that codes the exact location of an internal noxious stimulus. After all, the organism can do little to avoid it. Moving simply takes the affected body part with it. This mechanism underlies the harrowing nature of visceral and musculoskeletal pain, because there is nothing the patient can do to escape the pain. No matter where they turn the pain stays with them. The only purpose of the paleospino-thalamic pathway is to warn that the body is injured internally and to avoid stressing the affected part; but the disease itself cannot be avoided.

PAIN MODULATION

Within the brainstem are several centres that exert a descending inhibitory influence on nociceptive primary afferents and on second-order nociceptive neurones.[4] The principal centres are located in the periventricular grey matter, peri-aqueductal grey matter, the nucleus reticularis gigantocellularis, nucleus raphe magnus, the locus coeruleus and the lateral medullary reticular formation. Stimulation of these centres inhibits the transmission of nociceptive information between primary afferents and second-order neurones in the spinal cord. Axons projecting to the spinal cord arise from the nucleus raphe magnus, the locus coeruleus and the lateral medullary reticular formation. These centres themselves are controlled by the peri-aqueductal grey matter and the nucleus reticularis gigantocellularis (see Figure 3.7).

Descending axons pass along the dorsolateral funiculus of the spinal cord and reach all segments of the spinal cord, entering lamina II of the dorsal horn (see Figure 3.7). Here they exert excitatory and inhibitory influences on nociceptive neurones both directly and indirectly via islet cells and stalked cells in lamina II and inhibitory interneurones in lamina I (Figure 3.9).

Direct connections inhibit marginal neurones and excitatory stalked cells. Indirect connections excite inhibitory islet cells that postsynaptically inhibit marginal neurones. Other axons excite inhibitory interneurones in lamina I that presynaptically inhibit primary afferents reaching the dendrites of marginal neurones (Figure 3.9).

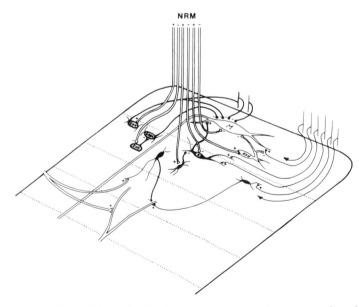

Figure 3.9   Descending inhibition of nociceptive neurones. Descending neurones from the nucleus raphe magnus (NRM) establish various connections in the dorsal horn. Some axons directly inhibit marginal neurones. Others stimulate inhibitory neurones in lamina I that use dynorphin (DYN) as their transmitter to inhibit marginal neurones and primary nociceptive afferents. Descending inhibitory axons inhibit excitatory stalked cells (s) that relay primary nociceptive information to marginal neurones. Excitatory axons stimulate islet cells that inhibit stalked cells. Excitatory axons inhibit enkephalinergic interneurones (ENK) that control other inhibitory neurones (GABA) that use gamma-aminobutyric acid as their transmitter to control marginal neurones. Excitatory axons stimulate enkephalinergic interneurones that inhibit marginal neurones.

These connections involve a variety of transmitter substances. Descending axons from the nucleus raphe magnus use serotonin as their transmitter substance. Axons from the lateral medullary reticular formation use noradrenaline. Inhibitory neurones in lamina II use enkephalin and GABA. Inhibitory neurones in lamina I use dynorphin.

Endorphins and enkephalins

A variety of transmitter substances occur in the central nervous system that belong to a class of chemicals known as endorphins and enkephalins.[4] They are all polypeptides. Endorphin itself is a large molecule consisting of 31 amino acids. It is found largely in the pituitary gland where it is synthesized from pro-opiomelanocortin, the precursor for ACTH and MSH. Endorphin has analgesic properties but it is not a conventional neurotransmitter. It is released largely in conjunction with ACTH into the blood stream and presumably has a role in providing analgesia during periods of major stress. It is therefore more a hormone than a transmitter substance.

The enkephalins are short chain peptides consisting of only 5 amino acids. The two main varieties are leucine–enkephalin and methionine–enkephalin. Their sequences are illustrated in Figure 3.10. What is striking is that the first four amino acids of enkephalin are identical to the terminal sequence of endorphin, and it is this set of amino acids that confers analgesic properties to both endorphin and the enkephalins.

The enkephalins are transmitter substances. They are found in a variety of locations throughout the body and not just in the nociceptive system. They occur in the cerebral cortex, the basal ganglia, in sympathetic ganglia and in the myenteric plexus of the gut. In the nociceptive system they occur in the peri-aqueductal grey matter, the nucleus raphe magnus and in lamina II of the dorsal horn. Whatever their location, enkephalins act as inhibitory transmitter substances causing hyperpolarization of postsynaptic membranes.

Dynorphin is another peptide consisting of 17 amino acids, the terminal five of which is the same sequence as leucine–enkephalin. It too is inhibitory and has analgesic properties.

β-endorphin	Tyr-Gly-Gly-Phe-Met-Thr-Ser-Glu-Lys-Ser- Gln-Thr-Pro-Leu-Val- Thr-Leu-Phe-Lys-Asn- Ala-Ile-Val-Lys-Asn- Ala-His-Lys-Gln-Gln
Met–enkephalin	Tyr-Gly-Gly-Phe-Met
Leu–enkephalin	Tyr-Gly-Gly-Phe-Leu
Dynorphin	Tyr-Gly-Gly-Phe-Leu-Arg-Arg-Ile-Arg-Pro- Lys-Leu-Lys-Tyr-Asp- Asn-Gln

Figure 3.10   The amino acid sequences of the endorphins and enkephalins. Note the similarity in the first four and five amino acids.

The analgesic properties of the enkephalins and dynorphin arise because they are transmitter substances in the nociceptive system. They occur at sites where the inhibitory influences they exert inhibit pain perception or block nociceptive transmission. The naturally occurring and synthetic narcotics such as morphine and pethidine act as analgesics because when they enter the central nervous system they mimic the inhibitory effects of enkephalins and dynorphin and thereby block 'pain' (see below).

Sensory discrimination

The existence of descending pathways capable of inhibiting nociception raises questions as to their natural function. It is tempting to believe that the body is equipped with its own analgesic system. However, this is manifestly not the case. Were it so, the body would automatically inhibit pain, but the existence of prolonged pain and chronic pain reveals that this is not the case. The function of the descending pathways is more sophisticated. It is part of the body's sensory discriminatory mechanism designed to filter out unwanted sensory information and to highlight information in which the brain might be interested.

At rest, the descending pathways are tonically active; they maintain a certain level of inhibition of the connections between primary afferent fibres and second-order neurones in the spinal cord.[8] When nociceptive information enters the spinal cord it competes for recognition with other sensory information reaching the brain. Activation of the spino-thalamic tract is not enough in this regard (Figure 3.11a). Unless the information is somehow highlighted it may be missed. In the case of a nociceptive signal, the body cannot afford to miss information about a threatening stimulus, so the central nervous system enhances the perception of the signal.

In addition to reaching the thalamus, nociceptive information uses spino-reticular axons or the collaterals of spino-thalamic axons to activate the nucleus reticularis gigantocellularis. This in turn activates the nucleus raphe magnus and the lateral medullary reticular formation either directly or via the peri-aqueductal grey matter. The net result is that the tonic influence of the descending pathways is modified. The modification is such that the tonic inhibition of the spinal cord's segments receiving the nociceptive information is reduced while adjoining segments are inhibited further. The effect is to highlight the incoming signal by enhancing its transmission while suppressing surrounding inputs (Figure 3.11b). This latter effect suppresses information that constitutes background noise and which might otherwise obscure perception of the incoming signal. This entire process is an example of centre-surround inhibition, a process used in many other sensory systems such as the eye and the posterior columns to enhance a desired signal by surrounding it with a field of inhibition. In other words, the central nervous system enhances the contrast of the incoming signal.

The role of the descending pathways is therefore to subserve sensory discrimination and not to provide the body with analgesia. It is only fortuitous that various therapeutic techniques can tap into this system and provide analgesia. They do this not by recruiting or mimicking the body's own pain relief system but by artificially disturbing its sensory discrimination system (see below).

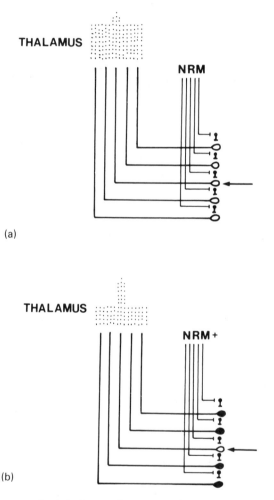

Figure 3.11 The sensory discrimination mechanism. (a) The diagram schematically illustrates an array of nociceptive neurones (open circles) whose axons project to the thalamus. These neurones are controlled by inhibitory interneurones (black circles) whose activity is controlled by the nucleus raphe magnus (NRM). A noxious stimulus delivered to the middle nociceptive neurone (arrow) evokes activity in the thalamus (dots), which is barely discernible amongst the background activity of surrounding neurones. (b) If the NRM inhibits all neurones surrounding the one transmitting the nociceptive signal, their activity in the thalamus is suppressed, and the nociceptive signal stands out in contrast.

Referred pain

Clinicians have laboured under a misconception about the perception of pain. They have an expectation that the patient suffering the pain should be able to indicate where it is coming from. Consequently, when referred pain occurs it suggests that something different, something special is happening.

Referred pain is pain perceived as arising in a location other than that of the source of pain. For example, pain may be perceived in the umbilical region when its source is in the appendix; pain may occur in the shoulder when its cause is blood irritating the abdominal surface of the diaphragm. In the case of musculoskeletal sources, pain may be felt in the buttock when its source is in the lumbar spine; pain may be felt in the head when its source is in the upper cervical spine. These seemingly bizarre associations invite an explanation. They imply that some special mechanism must be operating to generate referred pain, but the mechanism of referred pain has for many years remained elusive. Yet, the problem lies not in failing to find a mechanism, but in having a misconception about pain and its localization. The only form of pain that is felt locally is pain from skin. All other pain (from deep tissues) is referred pain to a greater or lesser extent.

When a noxious stimulus is delivered to skin it is accurately localized for two reasons. Firstly, the neospino-thalamic pathways that mediate cutaneous nociception are highly organized somatotopically. The pathways code for the location of the stimulus and the information is delivered to the VPL nucleus of the thalamus and the parietal lobe (see above). Secondly, very few external noxious stimuli selectively affect only nociceptors, and touch and pressure receptors are also activated by mechanical noxious stimuli. Impulses from these non-nociceptive receptors are delivered to the VPL nucleus of the thalamus by the posterior columns which again, are highly organized somatotopically, more so than the neospino-thalamic pathways. Therefore, the thalamus and brain receive two complementary sets of information that incorporate codes as to the location of the source of the noxious stimulus, whereby the brain registers virtually the exact point of stimulation on the skin. Indeed, the role of the posterior columns is such that without them the ability to identify the source of noxious cutaneous stimuli is reduced by 50%.

It is appropriate for cutaneous noxious stimuli to be well localized because they pose a threat to the organism. By identifying the exact source of stimulation the organism can take action to avoid the stimulus. The same does not apply for deep, internal noxious stimuli.

From a point of view of design, an organism does not need to accurately localize internal noxious stimuli because there is little it can do to avoid them; it cannot turn or run away; it cannot brush the stimulus away. All it requires is a general idea of the source of the pain so that it can avoid aggravating the pain. This is provided by the paleospino-thalamic pathway.

The spino–reticulo–thalamic connections of the paleospino-thalamic pathways provide information to the thalamus and brain that a noxious event is occurring. This establishes the appropriate alarm and concern. The localization they provide is relatively poor, but nevertheless is sufficient for the concerns of the organism. As a general rule, the spino-reticular system provides localization of the stimulus to within about one neural segment. Therefore, instead of being able to identify the exact point of stimulation, the brain registers the signal as arising from somewhere within those structures supplied by a certain spinal cord segment.

Therefore, for example, a patient may feel pain coming from the knee joint but cannot define that it arises specifically from the anterior cruciate ligament. As a rule, deep muscle pain is not perceived from the site of stimulation within the muscle but usually as if it is coming from the joint that the muscle moves. The deeper and more obscure and unfamiliar the source of pain, the less distinct its localization. The pain may be centred around the source but the boundaries of the pain are indistinct and extend beyond the limits of its source. This is not so much a characteristic of referred pain as it is of musculoskeletal pain and other deep pain in general, be it referred or not.

The reasons for this are twofold. Firstly, the spino-reticular system does not code for the location of deep noxious stimuli as well as the neospino-thalamic pathways do for cutaneous stimuli, but secondly, and moreover, deep pain does not involve the activation of non-nociceptive cutaneous receptors, which of their own accord provide localizing information to supplement that carried by the nociceptive pathways. In other words, deep pain is recognized as unpleasant but lacks accompanying information as to where it is actually coming from.

It is only in the case of joints that additional information about location may be provided. Proprioception travels along the posterior columns and is well organized somatotopically. Therefore, a patient will be able to localize joint pain because moving the joint aggravates the pain but whereas the nociceptive pathway provides information about the nature of the noxious stimulus, it is the posterior columns that provide parallel information as to its origin. In registering movement at the same time as aggravation of the pain, the brain is able to deduce that the structure that is moving is the source of the pain. This facility is less available for muscles and not at all for bones, because bones do not move and are not endowed with propriocep-tors. As a rule, the more a structure lacks cutaneous and proprioceptive innervation the less apparatus it has to localize pain, and the more indistinct the perception of noxious stimuli arising from it.

Patients have the ability to explore their injured parts. This can provide sensory cues. If the pain arises from a deep structure near to skin, palpation of the part evokes tenderness. Consequently, the source of pain may be localized on the basis of localizing information from the skin that was palpated. By simultaneously receiving nociceptive information and touch information the brain is able to deduce that the source of the pain must be near the site that was touched. This facility is not available for deeper structures. No amount of touching skin aggravates the pain. The patient is frustrated in their exploration and is left with an unclear message as to the source of the pain.

Perhaps the most striking example in this regard is the pain of sphenoid sinusitis. The sphenoid sinus is the deepest portion of the head; it cannot be palpated, and anyone who has not studied anatomy would not know what the sphenoid sinus is let alone where it is. The bone does not move and so has no proprioceptive innervation. Consequently, the pain is poorly localized. The patient can only report that it is somewhere in the field of the trigeminal nerve, i.e. somewhere in the head. The complaint is therefore solely one of headache.

Similar remarks apply to varieties of spinal pain. If the pain stems from a deeply located joint that cannot be palpated, whose whereabouts are unfamiliar to the patient, and whose proprioceptive output is only one part of the diffuse proprioceptive output from the vertebral column, there simply is no means whereby the patient can specify the actual origin of the pain. The best they can report is that the pain arises somewhere from within the neural segment or segments that happen to innervate the affected joint.

In these terms, it is evident that no special mechanism operates for referred pain; it is simply a reflection of the lack of localizing information accompanying deep nociception in general. The deeper and more inaccessible the source of pain is, the more indistinct are noxious stimuli arising from it, and the more widespread is the complaint of pain. In such situations it is the role of the clinician to use a knowledge of anatomy to identify the possible sources of pain, and in this regard a knowledge of the segmental innervation of the body becomes paramount.

The segmental innervation in question is not that of the dermatomes. Deep pain is not perceived in the skin; it is perceived deeply. What is therefore required is a knowledge of the deep segmental innervation of the body. In the case of the trunk, the pattern is straightforward. Virtually the entire trunk is supplied by intercostal nerves. The body parts (muscles, bones, joints and ligaments) innervated by an intercostal nerve are those lying in or limiting its intercostal space (Figure 3.12). Therefore, tracing the T4 intercostal space for example, detects all the structures innervated by the T4 nerve.

The lower six intercostal nerves leave the chest and continue into the abdominal wall, but do so in parallel bands. The location of these bands is indicated by the direction of the respective ribs (although not the costal cartilages). Whereas the lower costal cartilages turn upwards towards the sternum, their respective ribs point downwards and forwards towards the abdominal wall. It is this same direction that the lower intercostal nerves follow. Therefore, the 10th rib points towards the umbilicus, whereupon the umbilical region is innervated by the T10 intercostal nerve. The 12th rib points downwards towards the groin, whereupon the suprapubic region is innervated by T12. The L1 segment is simply the next in sequence and innervates a band of tissue ending in the inguinal region (Figure 3.12).

Deep pain arising in any of the thoracic or L1 segments will project into the

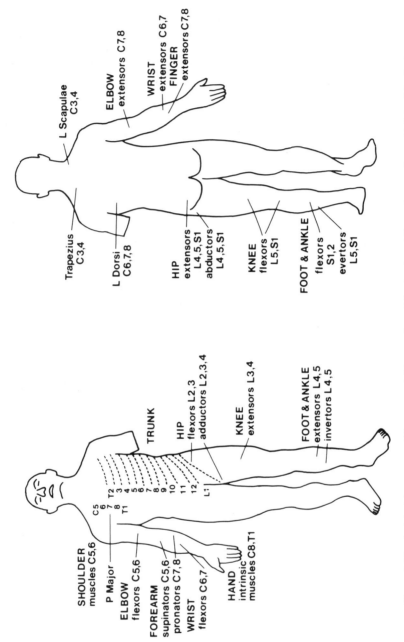

Figure 3.12   A pictorial summary of the segmental innervation of the trunk and the major muscles and joints of the limbs.

corresponding segment of the trunk wall as described above. Therefore, when a patient indicates pain in a particular trunk segment or segments, the diagnostic exercise requires first recognizing which segments are being indicated, and then a deduction of what other tissues are innervated by the same segment, and which therefore might be the actual source of pain. This may be any of the bones, ligaments, muscles or joints innervated by the segment including structures in the vertebral column as well as in the trunk wall, but in the case of trunk pain it also includes viscera. The interpretation of trunk pain therefore requires a knowledge of the segmental innervation of viscera.

The innervation of viscera is not constant, and is not accurately known, but sufficient knowledge is available to enable the formulation of worthwhile clinical rules (Figure 3.13).

All thoracic viscera are innervated by T1–T4 whereupon pain from any of these viscera will be perceived over the T1–T4 segments of the trunk, i.e. centrally over the chest or over the upper lateral chest wall.

The innervation of the abdominal viscera is essentially that of the alimentary tract. This tract is innervated sequentially from proximal to distal regardless of the actual location of the viscus. The thoracic oesophagus is innervated by T1–T4. The gastro-oesophageal junction is innervated by

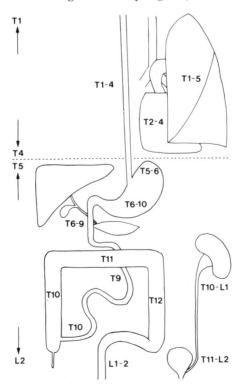

Figure 3.13   A pictorial summary of the segmental innervation of the thoracic and abdominal viscera.

T5–T6. The stomach itself is innervated by segments T6–T10. The duodenum and its derivatives (liver, bile ducts, gall bladder and pancreas) are innervated by T6–T9. The rest of the small intestines is innervated by segments T9–T10 with the T10 segment focusing over the terminal ilium and iliocaecal junction (hence the appendix belongs to T10). From proximal to distal, the ascending, transverse and descending colon is innervated by segments T10–T12, and the sigmoid colon and rectum is innervated by L1–L2. The kidneys, ureter and bladder are innervated by segments T10–L2 with the more proximal portions of the urinary tract receiving a higher segmental innervation (T10–L1) than the terminal ureter and bladder (T11–L2). Other pelvic organs are innervated by L1 and L2.

Given this pattern of innervation, it is evident that gastric pain will be perceived over segments T7±2 of the trunk, i.e. the epigastric region. In this case the organ in question happens to underlie the region of pain, but this is fortuitous. In the case of the transverse colon, this structure may lie in the epigastric region or umbilical region, but its innervation is about T11, and its pain is perceived lower in the abdomen and not in the epigastrium. Renal and ureteric pain occurs in the loin and travels to the inguinal region. This reflects the course and distribution of the T12 and L1 nerves in the trunk wall.

In the case of the limbs, the pattern of segmental innervation is topographically more complicated because in the course of the development of the limbs, what were once segmental bands in the embryo become twisted, mixed and convoluted in the formed limb. Nevertheless, a reasonable guide to the segmental innervation of limbs can be gained from a knowledge of myotomes, the segmental innervation of muscles. These are summarized in Figure 3.12.

From this guide it is evident that the 'top' of the shoulder is formed by trapezius (C3,4) and levator scapulae (C3,4). Note also that the diaphragm is innervated by the phrenic nerve (C3,4,5) and it is evident that pain from the diaphragm will be perceived in the shoulder, not because the shoulder is covered by the C3,4 dermatome (as is commonly taught) but because the deep tissues in this region belong to C3 and C4.

Pain stemming from cervical structures innervated by C5,6 will be referred to any of those portions of the upper limb that receive a similar innervation, i.e. the shoulder, the arm, the elbow and forearm. Lower cervical segments may refer pain to the arm, forearm, wrist and hand.

In the context it is noteworthy that the pectoralis major muscle is innervated by segments C5,6,7,8 and T1 but the muscle itself lies over the anterior chest wall. Therefore, quite apart from visceral pain referred to the chest wall (T1–T4), chest pain invites a consideration of neck pain from segments C5,6 and below referred to the pectoralis major.

In the lower limb it is evident that lower lumbar segments of the vertebral column share the same innervation as most of the muscles of the gluteal region and posterior thigh, and the buttock and thighs are the most common sites for referred pain from the lower lumbar spine. The L5 and S1 segments are also represented in the leg and ankle, and it is possible for low lumbar pain to be referred distally into the lower limb.

Neurogenic pain

Under normal circumstances, nociception involves the detection of damage or threatened damage in tissues innervated by the peripheral nervous system. However, in certain circumstances pain may be caused by nociceptive activity generated within nerves themselves. Peripheral tissues are not damaged and nerve endings are not stimulated, but nociceptive nerves are activated somewhere along their course.[9] An archetypal example of neurogenic pain is that arising from a traumatic or surgically-induced neuroma. When a peripheral nerve is cut or otherwise disrupted, within hours axons filaments sprout from the proximal end of the nerve. Their intention is to reach the distal stump of the nerve and to re-establish the connection. When this is successful the nerve regenerates and function is restored. However, the reconnection may be thwarted. The gap between the proximal and distal stumps may be too large because of retraction of the two ends, or the gap may become filled with scar. Under such conditions the axon sprouts grow aimlessly forming a tangle of filaments that constitute a lump on the nerve, the neuroma. The critical requirements for neuroma formation are that the continuity of axons must be broken and the epineural sheath must be breached. The latter allows axon sprouts to wander beyond the confines of the nerve. If the epineurium remains intact it confines the axon sprouts and directs them along and within the sheath to the distal stumps of the axons.

Not all neuromas become painful. Accurate figures are not available but perhaps some 10–15% become painful. The mechanisms for the pain are several. Neuromas are exquisitely sensitive to mechanical stimuli. They are also sensitive to circulating noradrenaline. Furthermore, they can become spontaneously active. Whatever the mechanism, activity in the damaged neurones is generated at the site of injury. However, although generated in the neuroma the activity is perceived as arising from the tissues that normally would have been innervated by the affected nerve. This is because the activity generated uses the same nerves and the same central connections that would be used by nociceptive information arising from peripheral tissue damage, but the brain has no way of determining that the origin of the information is actually from a neuroma and not from the peripheral tissues.

Other sources of neurogenic pain are the dorsal root ganglia. Impulses generated in a ganglion will be perceived as having arisen in those structures innervated by the spinal nerve to which the dorsal root ganglion belongs. In the case of spinal nerves that contribute to the brachial or lumbosacral plexuses the pain is perceived in the corresponding limb. Pain involving thoracic dorsal root ganglia is perceived around the chest wall or abdominal wall.

Ectopic activity in dorsal root ganglia may be evoked by a variety of means. Dorsal root ganglia are sensitive to mechanical stimulation, so that compression of a dorsal root ganglion by a disc prolapse, for example, may elicit bursts of neural activity. Alternatively, the ganglion may be pressed against a prolapse or an osteophyte as the nerve root moves during flexion or extension movements of the spine. Ganglion cells may be affected by inflammatory disorders, the archetypal example of which is herpes zoster or 'shingles'. A ganglion may become involved secondarily in inflammatory reactions to herniated disc material. If rendered ischaemic, dorsal root ganglion cells may fire spontaneously. This occurs in tabes dorsalis and when arteritis affects radicular vessels. It is also believed that nerve root ischaemia rather than simple mechanical pressure may be an operant factor in radiculopathies due to disc prolapse and foraminal stenosis.

To distinguish its origin, pain stemming from a dorsal root ganglion may be referred to as radicular pain, and it differs in quality from musculoskeletal pain. Radicular pain is lancinating in quality and is perceived as travelling through the tissues supplied by the affected nerve, typically along a narrow band. This contrasts with the constant, deep, dull aching pain of musculoskeletal origin. Furthermore, radicular pain is not purely nociceptive. Axons other than nociceptive axons are also triggered by the causative disorder. Consequently, the sensation has components beyond those purely of pain, but the patient is usually unaware of these. The massive discharge in many axons of different types is so alarming and overwhelmingly unpleasant that the entire sensation is described as painful, although close enquiry of the patient can reveal that the sensation is different from what they would ordinarily describe simply as pain.

Traditional wisdom maintains that nerve root compression can be a source of pain, but contemporary evidence refutes this. Mechanical stimulation of normal nerve roots does not evoke nociceptive activity. Only if nerve roots have previously been damaged are they capable of being painful. Otherwise, those pain states conventionally ascribed to nerve root compression are more likely to represent radicular pain due to some disorder affecting a dorsal root ganglia.

Central pain

Another form of neurogenic pain can arise from cells within the central nervous system.[10] It occurs when second- or third-order neurones in the nociceptive system lose their accustomed afferent input. This is reflected by the alternative nomenclature of 'deafferentation pain', but the term, 'central pain' is preferred because it specifies the origin of the nociceptive signal as being in the central nervous system.

When neurones lose their afferent input it appears that they undergo several physiological changes. They cease to maintain receptors, their membrane characteristics change, and they become spontaneously active and unresponsive to excitatory or inhibitory inputs. It is as if the neurone decides – 'if you are not going to talk to me why should I maintain receptors?'. In more physiological terms, it is more likely that transmitter substances released by peripheral nerves play a humoral role in maintaining receptors on second-order neurones. When peripheral nerves are destroyed or disconnected from the spinal cord, their transmitters are no longer supplied to the receptors and the maintaining, humoral affect is lost.

Central pain is the mechanism for phantom limb pain, paraplegic pain, the pain of brachial plexus avulsion and later post-herpetic neuralgia. The common feature of all these conditions is that peripheral nerves are destroyed or disconnected from the central nervous system, and cells in the central nervous system lose their accustomed input. Another example is thalamic syndrome, usually caused by an infarction, where it seems that neurones in the thalamic nuclei of the nociceptive system become spontaneously active as a result of loss of an accustomed inhibitory input or loss of an accustomed input from nociceptive tracts.

Central pain is notoriously difficult to treat because it is not due to a peripheral nociceptive input. It constitutes spontaneous activity within the central nervous system due to lack of peripheral input. Furthermore, once central neurones cease to maintain receptors they become unresponsive to pharmacological therapies that rely on mimicking the effect of inhibitory transmitter substances. In particular, if cells cease to maintain opiate receptors for enkephalin and dynorphin, they become unresponsive to morphine and other narcotics.

Of particular concern to physiotherapists and practitioners of musculoskeletal medicine is growing speculation that central pain may be the operant mechanism in conditions other than those involving major nerve injury. There is increasing evidence that some forms of joint pain to a greater or lesser degree involve changes in WDR neurones similar to those seen in central pain.[11] Therefore, some of the features of the pain of arthritis may not be nociceptive in the classical sense but involve an element of central pain. It behoves practitioners in this field to stay abreast of developments on the neurophysiology of joint pain, because it may transpire that classical interpretations and therapeutic approaches may need to be modified, focusing less on peripheral nociception and more on central mechanisms.

Serotoninopathy

There is increasing speculation that another form of central pain can occur. Nociceptive pathways are kept tonically suppressed by descending inputs from the brainstem and amongst these inputs are ones that use serotonin as their transmitter substance (see above). Although details have not been established, in principle it seems plausible that failure of serotonin to maintain this inhibition may create a disturbance in the sensory discrimination mechanism that controls the perception of pain such that an illusion of pain is produced. In physiological terms this pain is illusory because it does not represent actual nociceptive activity entering the central nervous system. Rather, the imbalance in the central nervous system creates a pattern of activity that the brain misinterprets as pain. To the patient, however, the pain is nonetheless real and distressing, if not severe, because even though there is no peripheral nociceptive input, the pattern generated within the nervous system is still real.

The basis for this disturbance appears to be some form of chronic or periodic disturbance of serotonin metabolism or that of a closely related substance. As levels of serotonin in neurones fall or as serotonin becomes less effective, its inhibitory influence on nociceptive neurones decreases, thereby decreasing the balance of tonic inhibitory effects on nociceptive pathways.

There is evidence to suggest that this mechanism underlies common complaints such as tension headache and migraine. In the case of migraine, it appears that headache occurs because of periodic failure of the locus coeruleus and raphe nuclei to suppress activity in the C2–C3 segments of the spinal cord. The vascular features of migraine are simply a parallel phenomenon, reflecting disturbances in the control of cerebral and cranial blood vessels by the locus coeruleus and raphe nuclei, but these vascular changes themselves do not cause the pain.[12] Serotonin is implicated in the pathogenesis of migraine not only by its presence in nociceptive pathways and the response of migraine to serotonergic drugs but also by the associated features of migraine such as mood disturbance, changes in appetite and vomiting, all of which processes involve serotonin or a related substance.

A similar relationship may apply for the mysterious yet seemingly common condition of fibromyalgia. This complaint is characterized by pain seemingly arising from muscle, associated with disturbances of mood, sleep and other functions. The common link is serotonin. Investigations of such patients have failed to reveal any primary disturbance in the affected muscles, yet the patients show disturbances of serotonin metabolism. A marker of this disorder is an increase in serotonin receptors on platelets.[13] Increases in receptors are a sign of chronic deficiency of the transmitter substance that acts on them, which suggests that fibromyalgia is not a peripheral nociceptive problem but a pain state due to deficiency of the set of serotonin within the central nervous system.

Similar inferences can be drawn about tension headache.[14] Investigations have failed to reveal consistent abnormalities of personality or muscle activity to justify ascribing tension headache to psychological problems or to chronic muscle contraction. Yet, there is mounting evidence that tension headache is associated with abnormalities of serotonin metabolism. Tension headache may therefore constitute one extreme of a spectrum of disorders that include fibromyalgia and migraine, which are caused by central metabolic disturbances and not by peripheral nociception.

Reflex sympathetic dystrophy

Apart from causing pain, tissue damage evokes a repair response. At the site of injury, this involves a cellular response that constitutes inflammation. Inflammatory cells such as neutrophils and macrophages invade the region to remove cellular debris. Cells are attracted to the site of injury by chemotactic factors, and their delivery is enhanced by vasodilation. At a local level, vasodilation is promoted by chemotactic factors and by substance P released from nociceptive nerve endings. Vasodilation is further promoted by the sympathetic nervous system, which increases blood flow to the injured region. However, in certain pain states this sympathetic response can be disturbed. For reasons not as yet understood, certain injuries evoke a bizarre reaction in the sympathetic nervous system consisting in its early phases of excessive vasodilatory response followed later by a severe vasoconstrictive response.

Pain problems involving the sympathetic nervous system are known by a variety of names, but the two most commonly used are reflex sympathetic dystrophy (RSD) and causalgia.[15,16] Causalgia typically occurs following a partial injury to a major peripheral nerve such as the median nerve or the sciatic nerve. In contrast, RSD typically follows some form of musculoskeletal injury, which itself may be relatively quite trivial, like fingers being jammed in a door. Otherwise, the two conditions are virtually identical in their subsequent manifestations and associated features (Table 3.1).

The patient complains of severe burning pain over a wide area of the affected limb or body part, associated with increased sensitivity to stimuli in the skin; the threshold for noxious stimuli is lowered and even innocuous

TABLE 3.1
Clinical features of causalgia and reflex sympathetic dystrophy

	Early	*Intermediate*	*Late*
SYMPATHETIC FEATURES			
Skin	Warm	Cold	Cold
	Red	Cyanotic	Pale
		Glazed	Smooth
Hair		Loss	Denuded
Nails		Brittle	Brittle
Subcutaneous	Oedema	Brawny	Atrophy
Joints	Swollen	Thick	Fibrous
	Tender	Stiff	Ankylosed
Muscles	Spasm	Wasting	Atrophic
Bones	Hyperaemic	Osteoporotic	Atrophic
NEUROLOGICAL FEATURES			
Burning pain	Hyperaesthesia	Hyperalgesia	Allodynia

stimuli become painful (allodynia). The affected limb exhibits vasodilation; it becomes red, hot and sweaty; muscles become tender and spastic; joints become swollen; and bones become hyperaemic especially towards their ends. Biopsies reveal proliferation of synovial cells and fibroblasts in the affected joints. This 'angry' phase of sympathetic activity lasts for some weeks or months only to be replaced by a cold, deathly vasoconstrictive phase. The limb becomes colder, blue and withered; muscles atrophy and become fibrous; joints stiffen; bones becomes osteoporotic; and trophic changes in the skin, hair and nails set in.

The pain in these conditions is most likely a form of central pain. In the case of causalgia the nerve injury causing deafferentation is obvious. In the case of RSD nerve injury has not been documented but could well be present in deep sensory nerves that innervated the original site of musculo-skeletal injury but which lacked a cutaneous distribution, and the nerve injury is not apparent on clinical testing. In some cases, RSD may occur after injury or diseases involving the central nervous system such as mul-tiple sclerosis or infarctions. In some cases it may follow ischaemic heart disease.

The associated features all seem to be mediated by the sympathetic nervous system. The cutaneous hypersensitivity is produced by noradrena-line released from sympathetic nerves sensitizing peripheral nociceptors and touch receptors, rendering them more easily and more strongly activa-ted. The changes in skin, muscle, joints and bones are due to excessive vasodilation in these tissues but may also reflect a direct trophic effect of sympathetic nerves. The later, atrophic features are the converse of this vasodilation.

The role of the sympathetic nervous system in these conditions is indicated by the fact that if performed early in the condition, regional sympathetic nerve blocks or sympathectomy can abolish the neurological sensitivity and vasodilatory features. However, in established cases the changes may be irreversible.

In these conditions, although the sympathetic nervous system is in-volved, it does not transmit the pain. The pain occurs independently and appears to involve WDR neurones in the spinal cord. However, the sympa-thetic nervous system perpetuates the pain by reinforcing the sensitized peripheral input to these neurones.

NOCICEPTIVE THERAPY

Nociception can be managed at several levels, either singly or in combi-nation. Therapy can be directed to the disease process, injury or cause of pain; to the transduction process; to the peripheral transmission of noci-ceptive information; to its central transmission; and to its modulation (Figure 3.14).

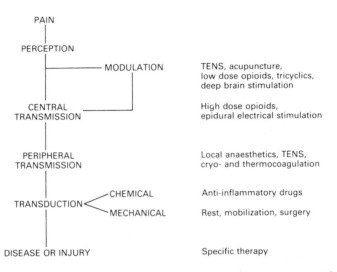

PAIN
|
PERCEPTION
|
|———————— MODULATION   TENS, acupuncture,
| | low dose opioids, tricyclics,
| | deep brain stimulation
| |
CENTRAL   ————————| High dose opioids,
TRANSMISSION   epidural electrical stimulation
|
|
PERIPHERAL   Local anaesthetics, TENS,
TRANSMISSION   cryo- and thermocoagulation
|
| ╱CHEMICAL   Anti-inflammatory drugs
TRANSDUCTION
| ╲MECHANICAL   Rest, mobilization, surgery
|
DISEASE OR INJURY   Specific therapy

Figure 3.14   A tabular summary of the process of nociception and pain perception and the sites at which different forms of therapy can be applied.

Disease process

Therapy directed at the disease process is the ideal, because if tissue damage can be resolved the trigger for nociception is removed. This can be done for relatively simple problems like sports injuries where in time tissue-healing occurs spontaneously. A therapist may then only have a role in attempting to accelerate or to optimize natural healing. In infectious diseases, a physician may assist the natural healing process by administering antibiotics, by lancing an abscess or by aspirating a septic joint. However, in many instances there may be no means by which the disease process can be reversed.

Some conditions lend themselves to radical excision, such as cholecystectomy for gall bladder pain, appendicectomy for appendicitis or excision of a peptic ulcer. However, such options apply largely for visceral diseases. In musculoskeletal disorders the option for radical surgery still applies but is not necessarily as universally successful as for visceral diseases. Painful joints may be debrided, fused, or excised and replaced. In the case of joints of the appendicular skeleton, joint surgery is reasonably successful but nevertheless is attended by limitations and significant risks of morbidity. In the case of spinal pain, arthrodesis has an unproven and controversial status.

Transduction

When a disease or injury cannot itself be addressed or can only be treated partially by medical or surgical means, the option arises of treating the pain at the transduction process. For chemical nociception, pharmacological tools are available. Drugs like aspirin and its congeners inhibit prostaglandin synthesis.[17] By eliminating prostaglandins they remove their facilitatory effect on other algogenic chemicals. This may be useful in reducing chemical nociception, but it does not eliminate it altogether. Corticosteroids have a more profound effect because they stabilize cell membranes thereby preventing not only the activation of phospholipase A (see above) but also inhibiting the release of lysosomal enzymes from inflammatory cells. However, because of their side effects, corticosteroids have a role restricted to only certain inflammatory pain conditions.

Apart from aspirin and corticosteroids, there are no drugs in general use whose action is on the transduction process of nociception. There is perhaps place for the development of drugs that might inhibit bradykinin or serotonin in the periphery but none is yet available. There are no drugs available that can inhibit the transduction of mechanical nociception. It is therefore futile to attempt to treat mechanical nociception with peripherally-acting drugs. Mechanical transduction can only be treated by correcting the mechanical abnormality triggering nociception.

Peripheral transmission

If nociception cannot be blocked at the transduction level there is scope for blocking its transmission along peripheral nerves. The principle is to stop conduction of action potentials. This can be done using drugs like local anaesthetics, but their usefulness is limited by their short duration of action. The major role of nerve blocks using local anaesthetics is diagnostic, to establish which particular nerve or nerves are mediating the pain, and other forms of therapy might be directed specifically at the nerves involved.[18] However, some pain problems, typically those involving muscles, can exhibit a lasting response to a single or to a series of injections of local anaesthetic. The temporary relief afforded by the local anaesthetic appears to provide a temporary respite during which the cause of pain mysteriously and spontaneously resolves. Such conditions are probably ones in which some minor mechanical disturbance causes muscles or parts of muscles to go into spasm and become painful. The spasm perpetuates the mechanical disturbance, but if the pain is relieved the spasm abates and the mechanical disturbance spontaneously resolves. If it does not, the pain reappears and is only temporarily responsive to local anaesthetics.

Transcutaneous electrical nerve stimulation (TENS) is another means of blocking peripheral transmission.[19] The exact mechanism by which TENS operates is still controversial. Both central and peripheral mechanisms may be involved. In the periphery, the dispute concerns whether nociceptive nerves are saturated or fatigued by the electrical stimulation rendering them incapable of nociception, or whether the electrical stimulus simply disturbs the frequency of discharges in peripheral axons of all types thereby decoding the nociceptive signal and replacing it with 'noise' that is perceived as a tingle instead of pain.

Whatever the exact mechanism, TENS can be a powerful tool for relieving pain. It works best when the electrodes can be placed over a nerve at a site between the source of pain and the central nervous system. It can still work in other situations by stimulating branches of the same spinal nerve as the one that innervates the source of pain, but if the electrodes are peripheral to the source of pain the analgesic potency is much less. It has also been demonstrated that TENS exerts good effects when applied over acupuncture points.

Peripheral nociception can be blocked surgically.[20] In the past, neurosurgeons were accustomed to transect nerves in an attempt to relieve nociception, but this approach is no longer favoured because of the risks of inducing neuroma formation or central pain, which essentially replaces the original pain with what is usually a worse form of pain. Instead of cutting peripheral nerves, they can be frozen or coagulated using needle-like electrodes inserted percutaneously onto the nerve.

Freezing nerves (cryo-neurotomy) turns intracellular water into ice which fractures the axon membranes. This interrupts conduction in the nerve for about six weeks or so, after which time the membrane heals and conduction is restored. Such treatment is not particularly useful for chronic pain but it can play a valuable role in conditions that are painful but destined to resolve in a matter of weeks. In the case of fractured ribs, intercostal nerves may be frozen to provide prolonged analgesia over the period it takes for the rib to heal.

Nerves can be thermo-coagulated using electrodes that transmit a radio frequency current (percutaneous radio frequency neurotomy). The electrode heats the nerve over a small area and denatures its proteins. Conduction remains blocked until the nerve repairs which, depending on the length of nerve coagulated, may take up to 12 months or more. This form of therapy has the prospect of providing valuable long-term analgesia, although it does not constitute a permanent cure for the pain.

Central transmission

The classical approach to stopping central transmission of nociceptive information is spino-thalamic tractotomy, a procedure in which the anterolateral funiculus is interrupted surgically, either by incision or by radio frequency neurotomy.[21] This form of therapy has been reserved largely for cancer pain in patients with a limited life-expectancy, because after about 12 months the effects of the operation wear off. The reasons for this are unclear, but hypotheses include nociceptive transmission along pathways other than the anterolateral funiculus (see above) or the onset of thalamic pain (see above).

Related to TENS is epidural electrical stimulation, a technique in which electrodes are introduced like a pacemaker into the epidural space, and used to stimulate the spinal cord.[22] The effect of the electrical current is to block nociceptive transmission. Its mechanism is unclear but may involve either the recruitment of descending inhibitory pathways or simply disturbing the frequency code of nociceptive information. Clinically, the effect is that of blanketing the region of pain with a sensation of tingling. This form of therapy has a limited efficacy for musculoskeletal pain but appears more efficacious for neurogenic pain.

Pain modulation

The strongest and most widely applied forms of nociceptive therapy operate by interfering with the sensory discrimination mechanism for pain. These include drug therapy, acupuncture, deep brain stimulation and possibly also TENS and dorsal column stimulation.

Morphine and other opiates are the strongest, most widely available analgesic drugs. They operate in either of two ways depending on the dose and route of administration. In conventional, small systemic doses (taken orally or by injection), morphine interferes with the supraspinal tonic inhibitory control for nociceptive transmission. It inhibits the tonic inhibition, thereby raising the background noise of sensory information entering the spinal cord. This does not block peripheral nociceptive input but obscures it. The nociceptive signal is still present but the brain cannot discriminate it from amongst the heightened level of background activity in the spinal cord, and pain ceases to be evoked.[8]

In higher doses, or if morphine is delivered by injection into the epidural or subarachnoid space, the drug reaches synapses in the spinal cord that involve enkephalin or dynorphin. Here, the morphine mimics the inhibitory action of these transmitter substances on nociceptive neurones. In essence it acts like a false transmitter substance. It is able to do so because although enkephalin and dynorphin are chemically different to morphine, the three-dimensional shape of morphine is virtually identical to that of enkephalin and dynorphin, i.e. their active radicals are spaced in exactly the same configuration. Consequently, enkephalin and dynorphin receptors on nociceptive neurones cannot distinguish whether it is morphine or enkephalin acting on them, and they proceed to hyperpolarize the neurone thereby blocking nociceptive transmission.

Tricyclic antidepressant drugs, like amitriptyline, are believed to offer an analgesic effect through their action on serotonin.[23] They block the re-uptake of serotonin after it has been released from descending inhibitory axons in the spinal cord. Consequently, the serotonin remains in the synapse longer and its inhibitory effect is prolonged. This appears to be the mechanism by which these drugs provide pain relief in conditions such as tension headache, migraine and fibromyalgia and why tricyclics can be a useful adjunct in the treatment of non-specific back pain.[23]

Research into the physiology of acupuncture indicates that it has an effect similar to that of morphine. Acupuncture operates largely by the effect of diffuse noxious inhibitory control. This is a process by which a new, noxious stimulus triggers the sensory discrimination mechanism to enhance its own input, but in so doing it inhibits any antecedent input to other spinal cord segments. In a patient with ongoing pain mediated by a particular spinal cord segment, the sensory discrimination mechanism will be operating to highlight that input. When an acupuncture signal is then delivered to some distant site, its input switches the sensory discrimination mechanism to enhance the acupuncture signal, and in so doing it suppresses all other activity in the spinal cord including the pre-existing pain.

Since it involves the sensory discrimination mechanism, acupuncture involves the serotonergic and noradrenergic descending pathways and enkephalinergic interneurones. Obliterating these constituents of the nociceptive system or blocking their transmitter substances abolishes the effects of acupuncture.

The supreme form of pain control is deep brain stimulation.[22] Electrodes may be introduced into the periventricular grey matter or peri-aqueductal grey matter and used to activate the entire descending inhibitory system of the brainstem. This provides a profound analgesia, but this form of therapy is not widely employed because of the delicate surgery involved, the risks of morbidity, the costs involved and the after-care required.

A variant of deep brain stimulation is a procedure in which the electrode is directed not at the peri-ventricular grey matter but into the thalamus near the terminals of the spino-thalamic tract. Stimulation at this site blocks the reception of nociceptive information by the thalamus. This form of stimulation is particularly useful for central pain.

Neurogenic pain

The treatment of neuromas is difficult and controversial. Over the years, various conservative and surgical therapies have been explored including hammering the neuroma, injecting it with alcohol, phenol, steroids and other sclerosants, surgical excision, burying the neuroma in bone, and encasing it in perspex or the like.[9,20] Burying the neuroma reduces its exposure to mechanical stimuli but does not address its spontaneous activity. Sclerosants can have only a temporary effect because the neuroma regenerates. Excising the neuroma simply invites a fresh neuroma at the site of transection of the nerve. Microsurgical ligation has been advocated and is logical in the sense that it aims to enclose the proximal stump of the nerve in its own epineural sheath to prevent outgrowth of axon filaments; but despite encouraging reports by its originators this procedure has not attracted widespread endorsement.

Radicular pain is difficult to treat conservatively. Ideally, the cause of pain should be eliminated if possible. Typically, this can be done surgically in the case of osteophytes, foraminal stenosis or disc prolapse. Chemonucleolysis constitutes an alternative, less invasive means of treating certain forms of disc prolapse. It operates by dissolving the prolapsed material. There is no evidence that traction, aimed at 'opening up' narrowed intervertebral foramina has any value in the treatment of radicular pain other than providing a temporary relief (if that) while the patient is under traction. Any foraminal enlargement provided by traction is summarily reversed upon resuming the upright posture.

If the cause of radiculopathy cannot be reversed, radicular pain is virtually intractable. Some authorities advocate the use of membrane stabilizing drugs like carbamazepine, dilantin or clonazepam in an attempt to reduce spontaneous activity in dorsal root ganglion cells, but the efficacy of these drugs is not reliable.

Radiculopathy due to auto-immune arteritis or inflammation is amenable to corticosteroids, which may be taken orally or injected epidurally on to the effected nerve root. In the case of post-herpetic neuralgia, amitryptyline is the only drug of the many that have been tried that has been shown to have any substantial beneficial effect.

Generally, physical therapy, acupuncture and electrical therapy (other than epidural stimulation) are notoriously ineffective for neurogenic pain.

Central pain

Central pain is difficult to treat because it is not caused by peripheral nociceptive input. Consequently, therapy directed at nociceptive transduction or peripheral transmission is futile. Furthermore, if the spontaneously active neurones in the central nervous system cease to maintain receptors, central pain will be unresponsive to any of the pain modulation therapies. If the nociceptive neurones lack opiate and serotonin receptors, they will be unresponsive to morphine, tricyclics, acupuncture and TENS. This is particularly relevant to physiotherapists who may be called upon to deal with patients suffering from central pain but who have been misdiagnosed as having some form of obscure musculoskeletal pain.

From a conservative perspective, distraction appears to be the single most effective treatment for central pain. The more the patient dwells on their pain, the more they suffer. If the patient can be distracted from their pain and gainfully employed, they become less aware of their pain and suffer less. Temporary relief from severe central pain can at times be achieved by intravenous infusion of lignocaine, and sometimes the relief can be long-lasting.

From a surgical perspective, central pain can be treated by epidural electrical stimulation or by deep brain stimulation if facilities for these procedures are available.[10,22] A relatively new surgical therapy is dorsal root entry zone lesions.[21] In conditions like brachial plexus avulsion and postherpetic neuralgia, where central pain stems from spontaneously active second-order neurones in the outer layers of the dorsal horn, the pain may be stopped by coagulating these cells with fine radio frequency electrodes introduced under direct vision into the spinal cord in the region of the entry zone of the dorsal root of the effected spinal cord segment.

Reflex sympathetic dystrophy

RSD and causalgia are miserable pain problems. Not only does the patient suffer severe burning pain, they cannot bear for the affected part to be touched, and meanwhile the musculoskeletal elements and skin undergo progressive trophic changes culminating in a withered, useless limb.

Physical methods are not applicable for the treatment of the actual pain (the mainstay are sympathetic blocks and drug therapy aimed at the sympathetic overactivity).[16] However, physical therapy has a crucial role to play in maintaining the integrity of the affected limb in order to prevent muscle atrophy and joint contractures, so that if the pain can in due course be relieved the patient is not left with a useless limb. The difficulty faced by the physiotherapist, however, is the inaccessibility of the limb. It is too sensitive and painful to be touched, even brushed, let alone manipulated. The cardinal recourse is to institute physical therapy while the affected limb is anaesthetized or under some form of sympathetic blockade. This reduces the sensitivity and allows the physiotherapist to deal with the muscle and joint problems. This combined approach requires coordination and cooperation between the physiotherapist and the medical team. When undertaken, the results are most rewarding, yet tragically, because of logistic difficulties or because of uninformed reluctance on the part of doctors, this type of therapy is all too often denied to patients.

SUMMARY

It is perhaps fitting to close with the preceding example of RSD, for no condition better illustrates the utility of a physiotherapist having insight into a pain problem. By understanding the mechanism of RSD, the physiotherapist should recognize that their techniques have little value for the pain itself, but the physiotherapist nonetheless has a crucial role in preserving the limb. Moreover, they should realize that medical assistance is required to institute local anaesthetic blockade to enable physiotherapy. Uninformed medical practitioners may not have considered this. Therefore, the responsibility may fall to the physiotherapist to become the patient's advocate. In order to do this effectively, the physiotherapist must be conversant with all of the dimensions of the problem and be able to explain this to their patient and to their colleagues. On the basis of a thorough knowledge of the physiology of pain, the physiotherapist should be able to indicate in a rational and logical manner what can and needs to be done, and not simply plead that something should be done by someone who knows more about the problem than themselves.

REFERENCES

1. Merskey H. ed. (1986). Classification of chronic pain. *Pain*, **Suppl. 3**, S1.
2. Turk D.C., Flor H. (1987). Pain > pain behaviours: the utility and limitations of the pain behaviour construct. *Pain*, **31**, 277.
3. Bonica J. (1990). General considerations of chronic pain. In *The Management of Pain* (Bonica J.J. ed.). Vol. 1, Philadelphia: Lea & Febiger.
4. Bonica J. (1990). Biochemistry and modulation of nociception and pain. In *The Management of Pain* (Bonica J.J. ed.). Vol. 1, Philadelphia: Lea & Febiger.
5. Bonica J. (1990). Anatomic and physiologic basis of nociception and pain. In *The Management of Pain* (Bonica J.J. ed.). Vol. 1, Philadelphia: Lea & Febiger.
6. Price D.D., Dubner R. (1977). Neurons that subserve the sensory-discriminative aspects of pain. *Pain*, **3**, 307.
7. Mountcastle V.B. (1968). Pain and temperature sensibilities. In *Medical Physiology* (Mountcastle V.B. ed.). 12th edn. St. Louis: Mosby, p. 1424.
8. Le Bars D., Dickenson A.H., Besson J.M. (1983). Opiate analgesia and descending control systems. *Advances in Pain Research and Therapy*. Vol. 5, New York: Raven Press.
9. Loeser J.D. (1990). Peripheral nerve disorders. In *The Management of Pain* (Bonica J.J. ed.). Vol. 1, Philadelphia: Lea & Febiger.
10. Tasker R.R. (1990). Pain resulting from central nervous system pathology (central pain). In *The Management of Pain* (Bonica J.J. ed.). Vol. 1, Philadelphia: Lea & Febiger.
11. Boivie J. (1989). On central pain and central pain mechanism. *Pain*, **38**, 121.
12. Lance J.W., Lambert G.A., Goadsby P.J., *et al.* (1983). Brainstem influences on the cephalic circulation: experimental data from cat and monkey of relevance to the mechanism of migraine. *Headache*, **23**, 258.
13. Moldofsky H. (1990). The contribution of sleep-wake physiology to fibromyalgia. In *Myofascial Pain and Fibromyalgia. Advances in Pain Research and Therapy* (Fricton J.R., Awad E.A. (eds.). Vol. 17, New York: Raven Press.
14. Olesen J., Langemark M. (1988). Mechanism of tension headache. In *Basic Mechanism of Headache* (Olesen J., Edvinsson L. eds.). Amsterdam: Elsevier.
15. Procacci P., Maresca M. (1987). Reflex sympathetic dystrophies and algodystrophies: historical and pathogenic considerations. *Pain*, **31**, 137.
16. Bonica J.J. (1990). Causalgia and other reflex sympathetic dystrophies. In *The Management of Pain* (Bonica J.J. ed.). Vol. 1, Philadelphia: Lea & Febiger.
17. Benedetti C., Butler S.H. (1990). Systemic analgesics. In *The Management of Pain* (Bonica J.J. ed.). Vol. II, Philadelphia: Lea & Febiger.
18. Bonica J.J., Buckley F.P. (1990). Regional analgesia with local anaesthetics. In *The Management of Pain* (Bonica J.J. ed.). Vol. II, Philadelphia: Lea & Febiger.
19. Sjolund B.H., Eriksson M., Loeser J.D. (1990). Transcutaneous and implanted electric stimulation of peripheral nerves. In *The Management of Pain* (Bonica J.J. ed.). Vol. II, Philadelphia: Lea & Febiger.
20. Loeser J.D., Sweet W.H., Tew J.M., *et al.* (1990). Neurosurgical operations involving peripheral nerves. In *The Management of Pain* (Bonica J.J. ed.). Vol. II, Philadelphia: Lea & Febiger.

21. Rosomoff H.L., Papo I., Loeser J.D. (1990). Neurosurgical operations on the spinal cord. In *The Management of Pain* (Bonica J.J. ed.). Vol. II, Philadelphia: Lea & Febiger.
22. Meyerson B.A. (1990). Electric stimulation of the spinal cord and brain. In *The Management of Pain* (Bonica J.J. ed.). Vol. II, Philadelphia: Lea & Febiger.
23. Monks R. (1990). Psychotropic drugs. In *The Management of Pain* (Bonica J.J. ed.). Vol. II, Philadelphia: Lea & Febiger.

Chapter 4

Mechanisms of Adaptation in the Joint

JOHN H. BLAND

INTRODUCTION

Optimum management in musculoskeletal physiotherapy is ideally based upon knowledge of gross and microscopic anatomy, the physiology of the structural elements involved, the mechanisms by which the connective tissue structures are normally maintained in optimal functional status and the regulation of the metabolic, biochemical and immunological processes in the tissues. The broad sweep of traumatic disorders, inflammatory diseases, genetically determined problems, birth defects, degenerative diseases – even malignant lesions – have a common theme, with its base in the emerging science of movement. To utilize all of the available scientific infrastructure in the practice of physiotherapy, one needs to be informed up to the cutting edge of knowledge regarding the normal regulation of form and function of the entire connective tissue system. It is the author's opinion that there is very much more known than is ever put into any practical use in physiotherapeutic practice.

The human joint is a remarkable piece of engineering, achieving as it does a balance of mobility and stability, a compromise which permits the joint an effective functional capacity. The joints are frequently affected in musculoskeletal conditions, either directly by trauma, infection or arthropathy, or indirectly as a consequence of immobilization, whether partial or complete. The involvement of the physiotherapist in the prevention or minimization of joint disorder and in the regaining of normal motion and function subsequent to such disorder is of considerable importance.

In this chapter, the normal physiologic and anatomic characteristics of the connective tissue, in particular those concerned with the articular structures, will be considered, along with the response of connective tissue to disease and trauma. The ability of the tissue to recover from such insults will be discussed, in particular with reference to the effects of immobilization and osteoarthritis, along with factors most likely to promote recovery.

HOMEOSTASIS IN BODY TISSUES

In the 19th century, Walter Cannon, an American physiologist, and Claude Bernard, the great French physiologist, evolved the general concepts of homeostasis of the internal environment in animal systems. In brief, this physiologic notion states that throughout the body's vascular and interstitial tissues and intracellular areas, there is a constancy of characteristics, including the biochemical and endocrine environments, the pH level, osmolarity, pressure and temperature. These 'constant' characteristics change within very narrow limits. For instance, the sodium concentration in extracelluar fluid remains very stable, despite the influence of many regulatory mechanisms such as dietary intake, and renal mechanism, sweat gland secretion and changes in the external environment. The temperature of the body is remarkably constant, fluctuating little more than one degree in a 24-hour period. The hydrogen ion concentration is guarded by buffering mechanisms in the pulmonary and renal systems.

The term homeostasis was coined late in the last century and implies that in a state of health a constant homeostasis is characteristic of animal systems. If these homeostatic states are disturbed, illness follows.

I propose here a related concept of homeostasis of all connective tissues in terms of their *physical properties*. Physical properties generally considered in connective tissues are tensile strength (as seen in tendon, the capsule of joints and in ligaments) and compressive strength (as demonstrated by hyaline cartilage). Since these are clearly physical properties, then what is responsible for maintaining the constancy of the physical properties of the materials that 'hold us together'?

The mechanism that maintains the physical properties of all connective tissues is *physical use*. This use provides mechanical stimulation in the form of stretching, pulling, jerking and compression. Connective tissue structures thrive, indeed absolutely require, mechanical stimulation by normal functional use and exercise. Skin, ligament, tendon, bone, cartilage and joint capsule undergo gross and striking changes if they are immobilized, even if they are only partially or relatively immobilized. Of course, overuse may change the physical properties of such tissues, although this is a relatively rare exception.

These ideas of the necessity and demand of connective tissues for continual mechanical stimulation to remain healthy and functional have clear cut and definite clinical implications for the day to day practice of the physiotherapist. In addition, the particular application of these principles to the joints, whether healthy or diseased, will be highly relevant to physiotherapy practice.

THE ARTICULAR ENVIRONMENT

Intra-articular temperature

The temperature in normal joints is between 32–34°C rather than the expected body temperature of 37°C.[1,2,3] Collagenase (an enzyme which is active in the breakdown of collagen) is produced by transformed synovial cells in joints of patients with rheumatoid arthritis and in very severe inflammatory osteoarthritis. These cells are maximally active at 37°C. Since the rate of enzymatic activity of collagenase is far less active at 32–34°C, it is desirable for the joint to be kept nearer 'normal' intra-articular temperature, rather than to show a raised temperature.

Harris and McCroskery[4] showed that increased articular temperatures of inflamed joints accelerate the rates of collagenolysis. Madreperla and colleagues[5] have described the induction of the synthesis of *heat-shock* proteins in joints at temperatures as low as 39°C and have linked these proteins to the initiation and progression of osteoarthritis. Effective treatment of such inflammation tends to lower intra-articular temperatures to the more nearly 'normal' levels, as shown by Hollander and Horvath.[1] We need more research on the subject of intra-articular temperature and its variations in the normal state and in inflammatory disease.

Intra-articular pressure

The pressure in normal joints is clearly subatmospheric, and this relative negative pressure creates a suction or strong vacuum. Joints are stabilized by this natural phenomenon. In the shoulder, the humeral head is situated firmly in the glenoid fossa, regardless of the range in which the arm is working, or whether it is used in lifting, pulling or twisting. The femoral head does not sublux during the swing-through phases of gait; it is held firmly in the acetabulum by the negative pressure in the hip joint. At rest and in motion the opposing surfaces of the articular cartilage are in close approximation at all times, that is to say, there is no actual space between the cartilage surfaces, rather a potential space. There is normally so little free fluid in the joint space that one cannot aspirate fluid with a needle. Muller[6] was the first to demonstrate pressures of −8 to −12 cmH₂O (−6 to −9 mmHg) in human knees. He also noted this subatmospheric pressure in the joints of anaesthetized dogs and in newly amputated limbs. This very important physical phenomenon has been amply confirmed in experiments by Levick and colleagues,[7,8] Treuhaft and colleagues[9] and Jayson and Dixon.[10–14] In the normal state, the negative (i.e. subatmospheric) pressure or vacuum in joints such as the knee increases with walking or other weight-bearing activities and with contraction of the periarticular muscles such as the quadriceps femoris (Figure 4.1).

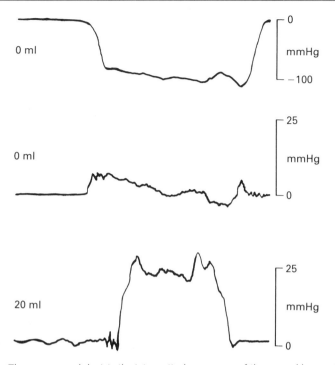

*Figure 4.1   The upper panel depicts the intra-articular pressure of the normal human knee at rest with contraction of the quadriceps muscle, i.e. subatmospheric pressure. The muscle contraction results in a marked decrease in pressure to −100 mmHg. The middle panel shows the effect of injection of 10 ml of normal saline. The quadriceps contraction causes a mild rise to a positive pressure. The lower panel demonstrates the effect of injection of 20 ml of normal saline. A rise in intra-articular pressure to 15 mmHg occurs. This positive pressure will just compress the arteriole–capillary–venule. Slightly higher pressure values would result in hypoxia–anoxia. (Reproduced with permission from Jayson M.I.V., Dixon A. St J. (1970). Ann. Rheum. Dis., **29**, 401.)*

A further interesting normal physiologic phenomenon is the following: with severe distraction of the joint the baseline pressure falls, reaching a level such that dissolved gases, e.g. nitrogen, come out of solution producing bubbles of gas in the joint – the well-known knuckle cracking. The bubble of gas and the joint distraction are readily demonstrated by serial radiographic study. This phenomenon is of more than passing interest. Unsworth and colleagues[15] showed experimentally that until a force equivalent to 10–16 kg was brought to bear, distracting the joint, there was no separation of the surfaces and no 'crack' or movement of nitrogen out of solution. At the point at which nitrogen moves out of solution, however, the force transmitted by the collateral ligaments suddenly increases, and the ligaments take over as the structures maintaining the integrity of the joint, replacing the force due to the negative intra-articular pressure (Figure 4.2). The gas remains as a bubble for about 20 minutes, and therefore a knuckle cracker must wait about 20 minutes between 'cracks'! Similar phenomena are seen in intervertebral discs, where the vacuum sign is well-known to radiologists (Figure 4.3).

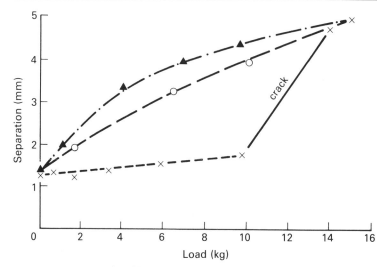

Figure 4.2   An experiment in knuckle-cracking. The ordinate depicts separation of the two surfaces of the middle metacarpophalangeal joint in millimetres measured by X-ray; the abscissa depicts the load or pull on the joint in kilogrammes. Note the 'X'-curve shows that 10 kg of force is required before the knuckle cracks and at that load the joint space separates abruptly. Until a bubble of gas appears in the joint and the pressure (negative prior to that point) becomes positive, ligaments and capsule take up the load. It is suggested that the normal vacuum holds the joint together until the pressure is so negative that nitrogen comes out of solution and does not return in solution for 20 minutes. With release of the load, the joint surfaces gradually return to baseline. (Reproduced with permission from Unsworth A., Doneson D., Wright V. (1971). Ann. Rheum. Dis., 30, 348.)

With the development of the Clark electrode in the early 1960s, the capability of measuring the intra-articular partial pressures of oxygen and carbon dioxide, the level of pH, lactic acid and glucose concentrations and temperature all became possible. Treuhaft and McCarty[9] and Falchuk and colleagues[16] independently studied these six parameters. A very low synovial fluid po_2 was found routinely in rheumatoid arthritis joint fluids, to everyone's surprise, often below 30 mmHg, and always accompanied by large decreases in pH and large increases in pco_2. At the same time, concentrations of lactate increased, reflecting the extreme acidotic state of the rheumatoid synovial fluid under its usual conditions. Much lesser magnitudes of these changes occur in the occasional extreme inflammation of osteoarthritis. These changes are strongly suggestive of a change in local tissue metabolism from principally aerobic to mainly anaerobic or glycolytic metabolism. These changes correlate with synovial cell proliferation, focal necrosis and severe obliterative microangiopathy.

In 1953, Ropes and Bauer[17] noted an inverse relationship between synovial fluid glucose and lactate levels. Glucose diffuses into cells, is phosphorylated and therefore is prevented from moving out of the joint space until it is metabolized to pyruvate. Pyruvate is, in turn, converted into lactate, which diffuses out of the cells into the tissue spaces and can be recycled. These data further confirm the grossly abnormal circulatory pH and gaseous changes that occur in the presence of an ordinary, relatively

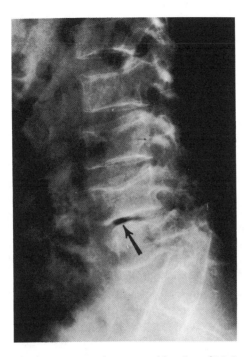

Figure 4.3   Lateral view of lumbar spine with anterior subluxation of L4–5 vertebrae. In the L5–S1 disc space is a bubble of nitrogen gas (arrowed).

small joint effusion. Clinically, the important issue is that these joints should not be exercised (except for simple range of motion exercises) because such activity would only increase the degree of acidosis, hypoxia, hypercapnia and necrosis of tissue.

Synovial diffusion

There is no thoroughly satisfactory means of studying synovial blood flow.[9] Xenon has been successfully used, but the variables are many. It is useful in studying the same patient serially to determine the blood flow in joints and such a study would be a valuable addition to the physiologic data we already have on joints.

The principal source of data on synovial diffusion capacity of various small molecules in water is the classical study of Simkin and Tizzorno.[18] Substances studied included tritiated water, urea, glucose, uric acid, phosphate, benzyl alcohol, creatinine and sucrose. Glucose, tritiated water and benzyl alcohol were all exceptions to the expected diffusion values for entry into and exit from the joint. All other molecules moved according to their diffusion coefficients as recorded in the standard literature. Perhaps because of synovial blood flow, the rate of egress of tritiated water was slower than that predicted from its diffusion coefficients alone. Lipid soluble benzyl alcohol had a rapid clearance rate, interpreted as indicating diffusion through synovial cell membranes as well as via the interstitial fluid pathway between the cells. Glucose was found to enter the joint rapidly and leave it very slowly.

THE DISORDERED JOINT

Joint effusion

According to Starling's law, the hydrostatic pressure at the arteriolar end of the capillary as it enters the capillary bed is about 22 mmHg. The tissue/inter-stitial fluid pressure tends to pull fluid out of the capillary, and at the venular end of the capillary the colloid osmotic pressure of serum proteins tends to pull fluid back into the circulation. Note that even small effusions of the order of 30 ml will raise the intra-articular pressure above 30 mmHg, thereby effectively compressing the small vessels and resulting in hypoxia. With a large effusion, *anoxia* occurs in these tissues. This matter is too little appreci-ated in clinical physiotherapy practice. The great majority of large effusions in rheumatoid joints, acute effusions in traumatized joints or intra-articular haemorrhage will result in hypoxic or anoxic joints (Figure 4.4). To heat such joints is to increase the blood flow in a situation in which the joint is getting all the blood it can handle, accentuating and accelerating an already serious degree of hypoxia for all the tissues of the joint.

In the presence of an effusion, the intra-articular pressure becomes rela-tively positive. When an effusion is present, the knee is at its most comfort-able position at about 30° of flexion, the position spontaneously adopted by patients with effusions. If quadriceps muscle contraction occurs and exten-sion follows, the intra-articular pressure become extremely high (500–1000 mmHg) and may even lead to rupture of the joint capsule. Such a rupture is more likely to occur in the acutely occurring effusion rather than the chronic effusion of osteoarthritis and rheumatoid arthritis, because the build up of fluid is not so rapid in these latter two conditions. The pressures of effusions are greatest at the knee in either full flexion or full extension and the lowest intra-articular pressure is generally at 30° of flexion.

The abnormal tissues in chronic synovitis are less compliant than the normal. Therefore, inflamed joint capsules in flexion or extension are at risk to rupture. Jayson and Dixon[10–12] observed a mean pressure of 802 mmHg in three rheumatoid arthritis patients injected with 100 ml of intra-articular fluid. In such instances one may see herniations through the capsule, rupture of the joint capsule and progressive distension of the synovial membrane (e.g. Baker's cyst in the popliteal fossa (Figure 4.5a). Such high pressures are active in the pathogenesis of the geodes or subchondral cysts in bone occurring in rheumatoid arthritis.[14] When rupture occurs, the syn-drome of *pseudothrombophlebitis* follows with a positive Homan's sign, tenderness of the calf and pitting oedema in the subcutaneous tissues. This

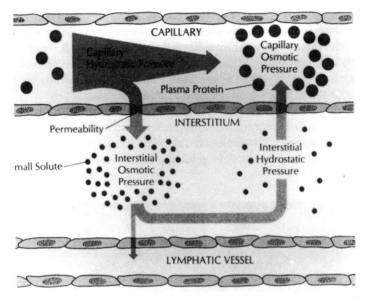

Figure 4.4   The hydrodynamics of arteriolar–capillary–postcapillary venule circulation. The capillary hydrostatic pressure is about 25 mmHg. As this falls, interstitial fluid osmotic pressure 'pulls' fluid into the extravascular spaces. As the venular end of the capillary is reached, the capillary osmotic pressure 'pulls' fluid back into the capillary and postcapillary venule. These vessels are soft and readily compressible. A positive pressure greater than 30 mmHg compresses the capillaries, compromising delivery of oxygen and consequent hypoxia or anoxia in and around the joint.

syndrome is frequently misdiagnosed as acute thrombophlebitis, resulting in unnecessary Doppler studies, venograms and administration of anticoagulants. Interestingly, one rarely sees true thrombophlebitis in the rheumatoid arthritic person. Note that a Baker's cyst is a normal bursa for the semimembranosus tendon and if and when it becomes distended or ruptured it may produce the syndrome described. More often, the cyst enlarges and may dissect down the calf, sometimes as far as the ankle in a massive, but unruptured cyst (Figure 4.5b).

Inflammation of the joint

In any joint with inflammation from any cause, there is always a parallel healing process actively engaged, little though it may be appreciated. During an episode in which gross inflammation is present, the tissues are being colonized by lymphocytes, plasma cells and polymorphonuclear cells. These invading cells migrate into the synovial space from post-capillary venules and plasma cells and are responsible for the production of enormous amounts of immunoglobin (40 mg/day); billions of lymphocytes are also dying there each day. This intense, active inflammation is tantamount to an Arthus reaction going on in the joint.

Simultaneously, even in the most intense cases of inflammation, healing is identifiably in progress. Macrophages are promptly involved in phagocytosing the debris of inflammation; angiogenesis, the formation of new capillaries, proceeds rapidly as messenger compounds are produced, stimulating capillary proliferation; fibroblastic synthesis and proliferation are also seen clearly as an actively developed healing process.

The important issue is to stop the inflammation and allow the normal healing process to dominate. Healing must occur by itself; no drug directly brings about healing. Nature heals the joint, drugs are merely ancillary to the natural healing process. The goal is to hasten the healing mechanisms;

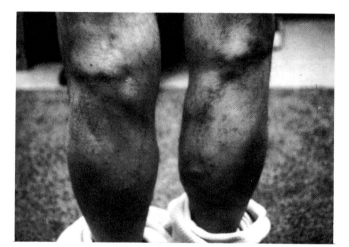

(a)

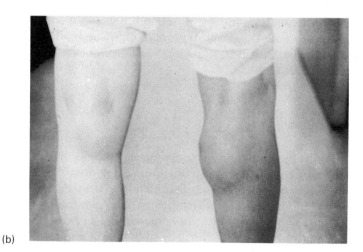

(b)

Figure 4.5   (a) Posterior aspect of the knees of a patient with rheumatoid arthritis, showing massive Baker's cysts that have herniated and dissected well down into the calf. (b) Baker's cysts. Note the massive cysts in the popliteal fossa with dissection medially down the calf.

exercise, in a form appropriate to the joint's condition, is mandatory to maintenance of normal physiologic mechanisms in all joint tissues.

Effect of motion on inflammation

Motion in a joint that is inflamed tends to accelerate the inflammation. Rest or immobilization (generally partial) helps to suppress inflammation. Gault and Spyker[19] carried out sequential studies into splinting of joints. A rest splint was applied to one wrist of a patient with rheumatoid arthritis, and the opposite wrist was used as a control. The splinted wrist showed progressively less inflammation, swelling and tenderness. This useful double-blind study did not, however, address the question of whether suppression of inflammation in many joints would occur if the whole body were to be immobilized.

Bland and Eddy[20] have shown that patients who have cerebral vascular accidents and subsequently develop osteoarthritis or rheumatoid arthritis do not get the inflammatory arthritis on the side of the paralysis, irrespective of how effective the rehabilitation of their neurological problem has been. The 'protection' against inflammation is not well understood – could it be related to the neurological manifestation? It is not necessarily a function of the immobilization, because many reported cases have shown preservation of both exercise and functional use. In an experimental model of joint inflammation, McCarty and colleagues[21] showed that the degree of inflammation caused by crystals of monosodium urate injected into dog knees was enhanced by a factor of 10 by simply passive movement of the joint.

Some motion, of course, must be maintained to avoid atrophy and contracture, but in general during an active and gross inflammatory process, range of motion exercises are all that are needed. The use of a skin-tight cast, which is removed serially as the inflammation and swelling subside, is especially effective as a form of immobilization in rheumatoid arthritis affecting the ankle, subtalar and midtarsal joints.

RESPONSES OF JOINTS TO IMMOBILIZATION

Wolff's law

Julius Wolff was a medical student at the University of Berlin in the 1860s.[22] He became intensely interested in bone and the changes that occur on altering the lines of physical stress. He evolved the Wolff's law over a period of 35 years. In its simplest form, the law states that bone will alter its size, shape and trabecular pattern in both the subchondral and cortical bone according to the lines of physical stress. *This implies that the normal size and shape of bone is regulated by normal lines of stress.* If there is a criticism of Wolff's law it would be that the law is far too restrictive. It really applies to *all* connective tissues, including ligament, muscle, bone, cartilage, joint structures, even fascia. I propose that Julius Wolff meant to say that this law could be adapted to all connective tissues.

We are familiar with Wolff's law of bone, but relatively few people are aware of the dependence of all connective tissues on physical stresses for maintenance of their homeostasis, a constancy of physical properties, to serve their function. This constancy is, in a sense, a physical counterpart to Bernard and Cannon's concept of homeostasis, the constancy of the internal environment.

The entire skeleton, as well as the whole connective tissue system surrounding it, can tolerate extremes of physical activity through its capability of perceiving and adapting to small and large changes in its functional cellular environment. Although the skeleton plays a critical role in mineral homeostasis of the body, it is primarily a structural entity, providing a framework and permitting locomotion. Bone mass increases with increasing physical activity and declines when the functional mechanical stimulus (e.g. exercise) is decreased or removed, such as is found during bedrest, lesser activity and microgravity conditions, or if the systematic stimulus is increased, such as through calcium deficiency or some endocrine diseases. The specific metabolic and physical parameters regulating as well as upsetting physical homeostasis of the skeleton are not entirely clear.

Remodelling is an alteration of internal and external architecture of skeletal tissue in response to the imposition of normal (or abnormal) physical stress. It implies removal of tissue at some areas, while it is being synthesized and laid down as tissue elsewhere. First Johnson[23] and others since[24] have expanded the concept to include changes in the size and shape of joints with age and osteoarthritis.

The form and function of all connective tissues are a consequence of changes in their internal architecture by 'self ordered' mathematical rules.[22] Stated otherwise, the morphology constitutes a balance between structural requirements (i.e. strength) and the metabolic advantage inherent in tissue economy (i.e. lightness). Morphology is substantially influenced by mechanical function, so remodelling locally and throughout the skeleton is under the constant influence of the degree and distribution of functional strains generated within the bone.[25] It is true that the length and general shape of bones and connective tissues are genetically determined, independent of function, but specific size and shape variations are due to functional load bearing. This life-long mechanical stimulation is mainly responsible for osteogenic or positive control of bone mass.

Although most of the studies made on the cellular basis of Wolff's law have been carried out on bone, there is more than enough of a body of literature to apply the principles to other connective tissues. Studies show a wholly predictable response of bone to whole body exercise[26] and to specifically localized exercise.[27,28] Bone changes resulting from space flight have been documented,[29–31] as has osteopenia adjacent to fracture plates[32] and to prosthetic components of various design.[33] There is strong experimental evidence of the great capability of all connective tissue systems (although most evidence to date concerns bone) to adapt to mechanical stimulation and its absence, confirming clinical observations such as in selective osteo-tomy, artificially applied loading and immobilization.[34-38] The wholly secure conclusion of these studies is that an increase in the level of physical activity (exercise) predictably leads to an increase in bone mass. Alternatively, a reduction in exercise activity leads to a decrease in bone mass.

Joint contracture

Although rheumatologists, orthopaedists and physiotherapists are fre-quently defeated by flexion contractures and the failure of healing of the connective tissues in trauma, rheumatoid arthritis and osteoarthritis, there is a reluctance to use the normal physiologic mechanisms available with regard to connective tissues and physical stresses in the therapeutic and management processes.[39]

There are protean and grossly destructive events imposed on any connec-tive tissue by total immobilization. Equally important is the gross disparity between the rapidity of onset and its protracted damage and the painfully extended and gradual recovery from this 'stress deprivation'. Table 4.1 lists in order the events, all of which are quite predictable, that occur in and about

---

TABLE 4.1

Effects of immobilization (stress deprivation) on synovial joint structures

---

SYNOVIAL JOINT SPACE
Fibro-fatty connective tissue proliferations – pannus

CARTILAGE
Fibro-fatty pannus proliferation over cartilage surface
Synovial cell proliferation – pannus
Ulceration at pressure contact points producing full thickness ulceration

ADHESIONS AS FIBRO-FATTY AND SYNOVIAL PANNUS MATURE
Tears at attachment sites: synovium; capsule; and cartilage
New planes of motion
Enzymatic degradation (by collagenases and proteoglycanases) of surrounding connective
    tissues

LIGAMENTS
Increased rate of synthesis of collagen
Loss of parallelism of collagen fibres
Increasing randomization in the pattern of deposition of collagen
Marked osteoclastic resorption at the ligament/bone attachment site
Interruption to collagen fibre continuity
Retention of attachment of periosteum only (after about 8 weeks)

TABLE 4.2
Effects of stress deprivation and exercise on synovial joints

STRESS DEPRIVATION

Increased rates of collagen and proteoglycan synthesis; increased randomized deposition of collagen fibrils and fibres; rapid, dramatic loss of tensile strength (8 to 12 weeks)

Fibro-fatty tissue proliferation over the cartilage surface, with pannus formation; ulceration of cartilage; tissue destruction; adhesions and flexion contractures

Osteoclastic destruction entheses

Recovery and healing incomplete after one year

EXERCISE

Slow increase in the tensile strength and hypertrophy of tendons and ligaments (taking at least one year to occur; changes in ligaments less marked)

Major energy input required for minimal change

Tendon repair markedly improved by early passive motion; intrinsic repair facilitated; rapid increase in tensile strength; no adhesions formed

joints that are totally immobilized. Table 4.2 is a summary of these events and the very slow recovery. The clinical issue is that the damage occurs very rapidly and the repair is extremely slow and less predictable, a heavy price to pay for immobilization – total, partial or even relative immobilization.

Strangely, ligaments in and around a totally immobilized joint demonstrate an *increase* in the production of collagen and proteoglycans by fibroblasts, quite contrary to predictions before experiments conducted in this area. Unfortunately however, the tropocollagen synthesized and exported to the extracellular space fails to aggregate in a parallel array. Instead, the molecules aggregate wholly at random, resulting in a disorganized tissue with bonding occurring equally at random. Such a tissue cannot have the high tensile strength required for ligamentous function. Figure 4.6 illustrates the normal parallel array of the anterior cruciate ligament in a normal exercising rabbit (left panel) and, in the right panel, a cruciate ligament from a totally immobilized rabbit knee showing the resultant random arrangement of the fibres, completely lacking the normal collagen parallel array, and therefore lacking the tensile strength required for normal function. In a sense, this is a new and very poorly functioning substitute for the normal tissue.

Equally unrecognized is the concept of partial or relative immobilization and their effects on the joint. If total immobilization is so grossly destructive, then it is reasonable to suspect that similar effects may arise from partial immobilization. The result of a decrease in physical activity combined with a reluctance to maintain a level of fitness with increasing chronological age and a general failure to appreciate the dependence on physical stresses for continued functional integrity will lead to the degradation of all connective tissues; in a word, retirement – both intellectual and physical.

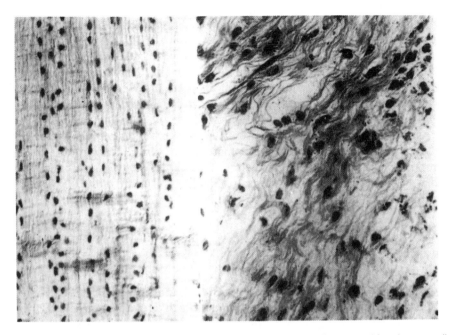

Figure 4.6   Left – section of a normal cruciate ligament from a normally active rabbit, showing cell nuclei and collagen fibres in a parallel array. Right – Cruciate ligament from a rabbit's knee immobilized for 10 weeks. Note the random distribution of collagen with loss of parallel array and hence marked changes in the physical properties of the ligament. (Reproduced with permission from Akeson S.H., Woo S.L.-Y., Amiel D. (1984). Rehabilitation of the Injured Knee (Hunter L. Y., Funk F. J. eds.). St Louis: CV Mosby.)

THE OSTEOARTHRITIC JOINT

Osteoarthritis is a very common, slowly progressive disorder occurring late in life, principally involving weight-bearing peripheral and axial articulations. Clinically, the disorder is characterized by pain, deformity, limitation of motion and, usually, slowly progressive joint destruction and consequent disability. Osteoarthritis has the greatest morbidity of all diseases. It is diagnosable in 35% of the knees of people as early as age 30, becoming almost universal in people over 70. At least 85% of people aged 70 to 79 years have diagnosable osteoarthritis.[40] Compromise of both structure and function is common, significantly interfering with quality of life.

The disease begins as a focal lesion, characterized by localized erosion of the surface of the hyaline cartilage (fibrillation) with shredding and loss of surface continuity. Chondrocyte clones form by mitosis and are associated with a marked increase in the rates of synthesis of type II collagen and proteoglycans, as well as increased levels of the enzymes concerned with the synthesis and self-contained remodelling of hyaline cartilage. These degradative enzymes appear in tissue as well as joint fluid. The subchondral bone osteoblasts increase their rate of synthesis, thickening and increasing the density of the subchondral bone. The entire end of the bone is remodelled, with the formation of marginal osteophytes, and the appearance of subarticular pseudocysts. Virtually all the tissues in and around the involved joint become thicker, hyperplastic and hypertrophic, with increasing clinical deformity. Not only are cartilage and bone grossly remodelled, but also capsule, tendon, ligament, and even muscle become hypertrophic, limiting motion. Loss of congruity of the articular surface and increasing mechanical instability follows.

Joint effusions occur, usually with nearly normal synovial fluid characteristics, synovial proliferation, inflammation, swelling and a gradually increasing loss of function. The patient is more likely to report pain than stiffness. The worst dysfunction in specific joints is found in the cervical and lumbar spine, the hips, knees and thumbs.

Myths and misconceptions

Table 4.3 lists the most common myths and misconceptions in osteoarthritis. Table 4.4 compares the consequences in joints of ageing versus osteoarthritis.[41] Hyaline cartilage can regenerate itself, although not in the form of the primary normal hyaline cartilage.[42] It is clear from contemporary evidence that chondrocytes can replicate themselves, and the wear and tear theory, based on the assumption that they cannot, is no longer valid.[43,44] Osteoarthritic lesions do not necessarily and inevitably progress with increasing disability and morbidity, and there is evidence that the condition can be arrested and reversed.

TABLE 4.3
Myths and misconceptions in osteoarthritis

---

1. A consequence of chronologic age
2. Secondary to wear and tear
3. Dogma – cartilage cannot heal and repair itself
4. Chondrocytes are effete cells, which cannot replicate or change rates of synthesis and degradation of cartilage macromolecules

---

Cell biology of osteoarthritis

Table 4.5 lists four events that occur in the course of osteoarthritis. This may follow a rapid or extremely gradual course over many years. First, a change in the microenvironment of the chondrocytes in the hyaline cartilage, most notably a decrease in the concentration of proteoglycans, is presumed to be the basis of the problem; the mechanisms by which this occurs are probably multiple and varied.

TABLE 4.4
Osteoarthritis versus ageing

Osteoarthritis	Ageing
Highly anabolic and synthetic process	Normal metabolism
Enzymatic destruction of hard tissue	Normal enzymatic modelling
Remodelling all tissues about joint (articular and periarticular)	Cartilage changes only
Chondrocyte mitosis	No mitosis
Intense increased synthesis of collagen and proteoglycan	Normal rates synthesis, collagen and proteoglycan
Increased water content cartilage	No change
Fibrillation, focal and progressive at weight-bearing sites	Fibrillation non-progressive, non-weight-bearing sites
Eburnation, ivory-like	No eburnation
Osteophytes occur with other changes	Osteophytes only with excessive use
No increased collagen cross-links	Increased collagen cross-links
Inflammation	No inflammation
No pigment – cartilage	Pigment – cartilage

Second, the cells located at the junction of the hyaline cartilage and the bone begin to overproduce a very distinctive type of bone. It is presumed that this is mediated by intercellular communication between the chondrocyte and the osteoblastic cells of the subchondral bone. With increased hyaline cartilage synthesis comes increased stiffness and decreased compliance, resulting in subchondral microfractures. These microfractures heal with callus, further increasing the stiffness and loss of compliance. Further microfractures occur and further callus is deposited.

The third step is initiated by intercellular communication between the chondrocyte and the osteoblast of the subchondral bone and synovial cells. Synovial cells become metaplastic producing a mixture of connective tissues (called chondro-osteophytes), cartilage, and an amorphous connective tissue, mostly bone, coated with a layer of fibrocartilage.

Lastly, through the microclefts of subchondral bone fractures, synovial fluid is extruded into the bone marrow, with resultant osteoblastic and fibroblastic reaction and the formation of cysts filled with connective tissue

TABLE 4.5
Steps in the pathophysiology of osteoarthritis

1. Microenvironment of chondrocytes changes; chondrocytes mitose, produce clones that increase rates of all export products, proteoglycans, collagen, enzymes
2. Subchondral bone osteoblasts increase rates of synthesis; density of subchondral bone increases; stiffness increases, microfractures follow
3. Osteophytes form at periphery of joint (metaplastic synovial cells), active inflammatory process, synovitis
4. Pseudocysts form in trabecular bone below subchondral bone; increased volume and density of all articular and periarticular structures, capsules, tendons, ligaments, bones

debris. This whole process, proceeding at varying rates, results in the clinical and radiologic characteristics of osteoarthritis, which in the extreme is deforming, crippling and results in the very high morbidity rates referred to above.

Historical tradition has tended to associate osteoarthritis with old age in an inextricable relationship and has accepted as inevitable a progressive deterioration and disability. It is pertinent for us to question the assumptions surrounding this tradition in the light of current knowledge.

Is osteoarthritis reversible or arrestable?

Population studies in the United Kingdom[45,46] and in the United States[47] have shown that extensive *asymptomatic* damage to the articular cartilage may exist in any joint. Forman and colleagues[48] have shown that osteoarthritis may not be progressive in older people. It is possible that some cases of painful advanced disease may become painless or may become intermittent. Most certainly hyaline cartilage, the tissue involved primarily in osteoarthritis, remodels continually in three ways: removal of existing cartilage; formation of new cartilage; and endochondral calcification and ossification.

Studies have reported marked clinical improvement and radiologic recovery of osteoarthritic hips on a number of occasions without active treatment intervention.[49–52] Harrison and colleagues[53] observed healing and considered that, as a consequence of the destructive process, fibrous bone marrow might reach the surface of the bone and differentiate into servicable fibrocartilage. Such fibrocartilaginous tufts usually remain discrete, but sometimes become confluent, resurfacing the bone. Johnson[23] suggests that the chondrocytes that produce fibrocartilage, under the right conditions of use and mechanical stimulation, can differentiate into chondrocytes that synthesize hyaline cartilage. This magnitude of healing in osteoarthritis, while uncommon, occurs in a setting of unaided natural forces. It is important that the mechanisms responsible for such a repair should be identified in the hope of being able to induce healing.

Salter and colleagues[54] in a 5-year study of experimental osteoarthritis in rabbits, showed that continuous passive motion is a powerful stimulus for healing and regenerating of hyaline cartilage. The distal femoral cartilage of each rabbit was damaged with four, full thickness drill holes into the subchondral bone. One group was immobilized post-operatively. In these rabbits, bone healed very slowly and the cartilage not at all. Joint adhesion and pannus formation were seen and further cartilage degeneration followed.

Another group were allowed to run free. Bone healed a little faster and the cartilage defects were filled with fibrous tissue with imperfect healing. A third group were subjected to continuous passive motion to the joint (begun immediately post-operatively) and demonstrated healing by regeneration of true hyaline cartilage – a process of normal chondrogenesis. Six months later, this latter group of animals had normal cartilage, and the other two groups of animals had osteoarthritis.

Immobilization and osteoarthritis

Ely and Mensor[55] showed that an immobilized joint develops cartilage changes, which they thought were similar to those of osteoarthritis. Simple immobilization in a cast resulted in hyaline cartilage changes, fibrillation, flaking, fissuring and overall thinning. In osteoarthritis, the chondrocytes proliferate in clones or brood clusters, over-producing their macromolecular export products. In contrast, during immobilization the chondrocytes actually degenerate and die in a few weeks to a few months, suggesting that, according to Wolff's law, they require mechanical stimulation to survive. Osteoarthritis develops over years, whereas immobilization arthropathy is characterized by an acute onset and abrupt alteration of joint function.

Salter and Field[56] and Enneking and Horowitz[57] showed that immobilization for long periods resulted in the proliferation of fibrofatty tissue that finally fills the joint, appearing very much like the pannus of rheumatoid arthritis, ultimately eroding the cartilage by enzymatic degradation. Even slight movement of the joint slows this process. Compression in addition to immobilization accelerates this process.[58] Palmoski and Brandt[59] showed a marked fall in the rate of synthesis of proteoglycan after 3 weeks in immobilized dog knees. Proteoglycan aggregates (with hyaluronate) no longer formed in the cartilage and would not aggregate normally with exogenous hyaluronate. With a return of mechanical stimulation, the joint use and the proteoglycan aggregation capability of the joint surfaces returned to normal, a clinically significant observation. Increased water content and loss of aggregability of proteoglycans did not return to normal if the animals were subjected to vigorous exercise instead of normal weight bearing. Therefore, immobilized joint cartilage showing signs of an osteoarthritic lesion is vulnerable to loading, and excessive exercise at that point accelerates the osteoarthritic sequence of events.

Neuromuscular reflex mechanisms and shock loading

Hyaline cartilage failure is caused by either excessive loading of normal cartilage or physiologic loading of abnormal cartilage. Study of the mechanism of shock absorption is important to the understanding of the normal regulation of cartilage metabolism and the acceleration by relative overload in the process of developing osteoarthritis. Shock is absorbed mainly by

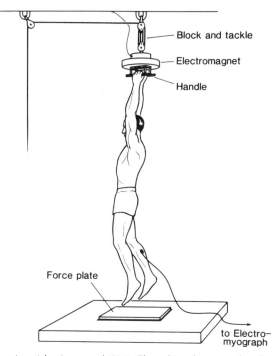

Figure 4.7   *An experiment by Jones and Watt. The subject hangs on handles connected to an electromagnet; a force platform is below him and an electromyograph is connected to his calf to record activity in the gastrocnemius/soleus muscle. The experiment was designed to study the 'functional stretch reflex', a mechanism for landing from unexpected falls. The subject falls, but does not know when he will be released. The force platform records the force and time of landing. The electromyograph records the time of muscle contraction. In landing from less than 9 cm to 13 cm, there was always a painful jolt. That is, the muscle contracted **after** the subject hit the force platform and the reflex failed to protect him. Landing from falls 15 cm to 18 cm and higher were not associated with pain or injury. Therefore, the reflex occurred prior to the subject hitting the force platform, taking up the impulsive force of the fall. (Reproduced with permission from Jones M.G., Watt D.G.D. (1971). J. Physiol., 219, 729.)*

three mechanisms: the hyaline cartilage itself, the expansion and knobbliness of the bone ends, and the neuromuscular reflex apparatus.

Hyaline cartilage is an excellent material for shock absorption, but is present in inadequate amounts. Bone, although not as effective in absorbing shock, is present in sufficient quantities to contribute to shock absorption. The main mechanism of shock absorption is by reflexly controlled neuromuscular mechanisms – active lengthening of the muscle while maintaining its tension.

Jones and Watt[27] carried out a series of elegant yet simple experiments disclosing that young healthy subjects could sustain significant injury in an unexpected fall if the height of the fall was too short to allow normal reflex muscle contraction and absorption of shock. The experiment is described and illustrated in Figure 4.7. This experiment has pertinence to the management and perhaps the development of osteoarthritis. It points to the necessity of maintaining optimum musculoskeletal fitness at all ages. Contrary to the usual excessive rest programmes in osteoarthritis, it is proposed that well governed and practised exercise and mechanical stimulation of the involved tissue may aid in arrest or reversal of the process. I propose that damage to joints may occur in persons who have lost fitness to a significant degree, i.e. have suffered gradual damage to the joints due to ordinary but limited activities of daily living.

Life-style influences on the osteoarthritic joint

Since osteoarthritis becomes universal with increasing chronological age, it is proposed that a contributing force in the development of osteoarthritis is the general loss of fitness, reflex responses, osteoporosis and overall decrease in mobility and intellectual powers, a consequence of both intellectual and physical inactivity – *the retirement concept*. Proper attitudes towards perceived health, body image, physical activity, anxiety and life satisfaction have been studied in men and women aged 60 years of age and older.[60] Virtually all measures of fitness improved by 83% over a 14-week endurance training programme, including maximal oxygen uptake, blood pressure, pulse rate, general sense of musculoskeletal stiffness and general perception of well-being.[61]

Sophisticated application of knowledge of nutritional requirements in the aged and the osteoarthritic is of major importance.[62,63] Malabsorption is common, as is alcoholism, with consequent malnutrition in the aged. Restriction of food markedly increases the life-span of rats and other laboratory rodents, and delays the appearance or slows the progression of those diseases believed to limit the life span of the animal.[62] Physical activity in people over 55 prevents the slow bone loss and even increases bone accretion. In addition, rates of normal mineral loss in the ageing osteoarthritic individual can be favourably affected. The increased calcium intake requirement and assurance of its absorption are achievable goals in the management of osteoarthritis. Deficient vitamin C intake is common and may be associated with a defect in collagen proline hydroxylation in collagen synthesis, reflected in the known excretion of proline rich collagen derived peptides. This may compromise connective tissue repair, a necessity in osteoarthritis. Muscle, tendon, ligament and joint capsule respond with increased synthesis, and hence tensile and compressive strength increase at all ages.

THE PHYSIOTHERAPIST AND THE DISORDERED JOINT

The physiotherapist has a fundamental role in the management of the disordered joint. In fulfilling this role, attention should be given to a number of considerations, in particular, certain physiologic and anatomic principles:

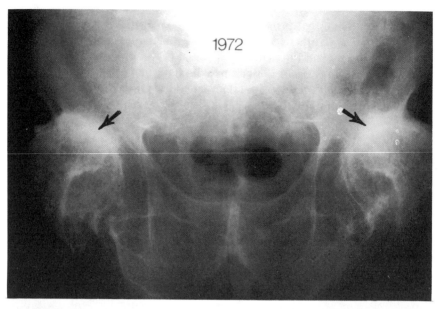

(a)

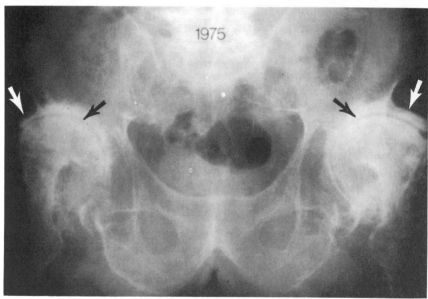

(b)

Figure 4.8   (a) An 85-year-old man bilateral osteoarthritis of both hips, first interviewed and examined in 1972. He was a 'bed and chair' patient and had been essentially immobilized for a decade. He was an athlete in younger years and continued until the age of 72, when his doctor advised rest, canes and acceptance of the inevitable. He was regarded as 'too old' for consideration for total hip replacement surgery. A programme of bed and chair exercises, home physical therapy, education into physiological principles, weight loss and general physical activity to tolerance was initiated. Note a medial joint space in both hips in 1972 and a seeming ankylosis in the superolateral area of both hips (arrowed). (b) Note the hip joints in 1975. There is a joint space detectable throughout. A dense shelf of bone has developed laterally in both sides and the femoral heads have undergone extensive remodelling (arrowed). He was now able to walk (with two canes) 1 km twice daily and had become pain free. The patient died in 1981 aged 94.

1 Cardiovascular effects of inactivity
2 Effects of partial and total immobilization on the connective tissues
3 Metabolic and endocrine changes with inactivity
4 The dangers of rest with respect to exercise responses
5 Psychosocial and chronophysiological effects of inactivity and immobilization
6 The impact of exercise after inactivity regarding diminished work capacity
7 The need for control of posture and the implications of loss of such control
8 The need to encourage controlled relaxation.

The case studies illustrate clinical examples of the practical applications of some of these principles (Figures 4.8, 4.9, 4.10 and 4.11).

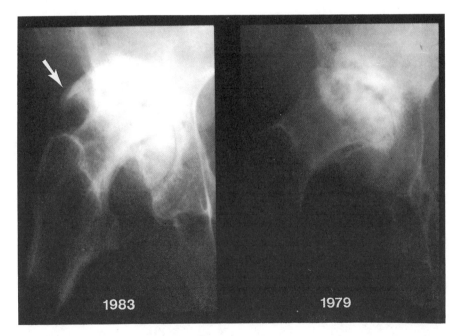

Figure 4.9   A 63-year-old woman with osteoarthritis of her right hip. Right panel – shows the joint in 1979 when it greatly limited her ambulation and was severely painful. Note the extensive loss of bone substance, large subchondral cysts, marked moulding of the femoral head, dense subchondral bone sclerosis and severe joint space narrowing and irregularity. Left panel – shows her hip in 1983 after a programme of planned exercise, rest, passive stretching of the joint and education in the principles pertaining to joint rehabilitation. Note the extensive remodelling, rounding of the femoral head and much new bone formation. A large shelf of bone now projects from the upper lateral acetabulum (arrow). There is a joint space, albeit thin, throughout. Subchondral cysts have been remodelled and new, smaller ones have appeared. There is much more overall sclerosis of bone throughout the femoral head. The patient was not pain-free, but found her limitations acceptable and preferable to total hip replacement. Her overall function and scope of physical activity were greatly improved.

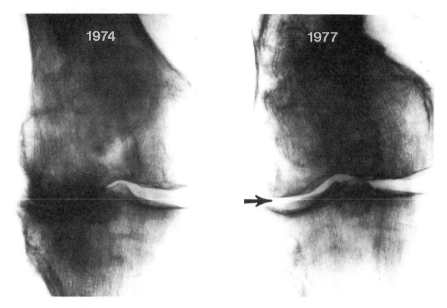

Figure 4.10 Left panel – a 66-year-old man with severe, very symptomatic uni-compartmental osteoarthritis of the knee. The patient had a successful osteotomy; note the deformity of the femoral shaft at the top of the picture. Right panel – taken 3 years after the operation, when the patient was asymptomatic, pain free and had developed a normal weight-bearing joint space (arrowed). Note the extensive remodelling of the dense, 'osteoarthritic' subchondral bone in this view compared to the earlier film. Note that these are positive films, with dense bone appearing darker.

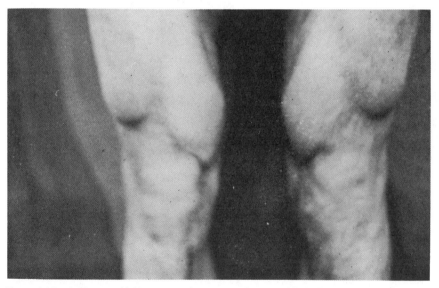

Figure 4.11 A 23-year-old Norwegian ski jumper who, in a fall, had torn his anterior cruciate ligaments (bilaterally) 4 years previously and had a positive drawer sign and unstable knees. This photograph was taken 3 years after initiation of an extensive programme of strengthening exercises for all the intra-articular and peri-articular structures. He had had no surgical procedures to the cruciate ligaments, so the status of these structures is not known with precision. In 18 months of a very vigorous programme his knees became stable and he competed as a ski jumper for two winter seasons. This photograph shows the hyperplasia of the connective tissues, muscle hypertrophy and increased tensile strength that can be attained by taking advantage of Wolff's law. The only regret is that no photographs exist of the pre-injury status.

REFERENCES

1. Hollander J.L., Horvath S.M. (1950). The influence of physical therapy procedures on the intra-articular temperature of normal and arthritic subjects. *Am. J. Med. Sci.*, **218**, 543.
2. Horvath S.M., Hollander J.L. (1949). Intra-articular temperature as a measure of joint reaction. *J. Clin. Invest.*, **28**, 469.
3. Hollander J.L., Stoner E.K., Brown E.M. (1951). Joint temperature measurement in the evaluation of anti-arthritic agents. *J. Clin. Invest.*, **30**, 701.
4. Harris E.D.J., McCroskery P.A. (1974). The influence of temperature and fibril stability on degradation of cartilage collagen by rheumatoid synovial collagenase. *N. Engl. J. Med.*, **290**, 1.
5. Madreperla S.A., Louwerenburg B., Mann R.W. (1985). Induction of heat-shock protein synthesis in chondrocytes at physiological temperatures. *J. Orthop. Res.*, **3**, 30.
6. Muller W. (1929). Uber den negativen luftdruck im gelenkraum. *Deutsche Zeitschrift fur Chirurgie*, **218**, 395.
7. Levick J.R. (1983). Joint pressure volume studies: their importance, design and interpretation. *J. Rheumatol.*, **10**, 353.
8. Levick J.R. (1983). Synovial fluid dynamics: the regulation of volume and pressure. In *Studies in Joint Disease* (Maroudas J., Holborow K. eds.). p. 153.
9. Treuhaft P.S., McCarty D.J. (1971). Synovial fluid pH, lactate, oxygen and carbon dioxide partial pressure in various joint diseases. *Arthritis Rheum.*, **14**, 475.
10. Jayson M.I.V., Dixon A. St J. (1970). Intra-articular pressure in rheumatoid arthritis of the knee: I. Pressure changes during passive joint distension. *Ann. Rheum. Dis.*, **29**, 261.
11. Jayson M.I.V., Dixon A. St J. (1970). Intra-articular pressure in rheumatoid arthritis of the knee: II. Effect of intra-articular pressure on blood circulation to the synovium. *Ann. Rheum. Dis.*, **29**, 266.
12. Jayson M.I.V., Dixon A. St J. (1970). Intra-articular pressure in rheumatoid arthritis of the knee: III. Pressure changes during joint use. *Ann. Rheum. Dis.*, **29**, 401.
13. Jayson M.I.V., Dixon A. St J. (1970). Valvular mechanisms in juxta-articular cysts. *Ann. Rheum. Dis.*, **29**, 415.
14. Jayson M.I.V., Dixon A. St J. (1970). Unusual geodes ('bone cysts') in rheumatoid arthritis. *Ann. Rheum. Dis.*, **31**, 174.
15. Unsworth A., Doneson D., Wright V. (1971). 'Cracking joints': a bioengineering study of cavitation in the metacarpophalangeal joint. *Ann. Rheum. Dis.*, **30**, 348.
16. Falchuk K.H., Goetzl E.J., Kulka J.P. (1970). Respiratory gases of synovial fluids: an approach to synovial tissue circulatory-metabolic imbalance on rheumatoid arthritis. *Am. Med. J.*, **49**, 223.
17. Ropes M.W., Bauer W. (1953). *Synovial fluid changes in joint disease*. Boston: Harvard University Press.
18. Simkin P.A., Pizzorro J.E. (1974). Transynovial exchange of small molecules in normal human subjects. *J. Appl. Physiol.*, **36**, 581.
19. Gault S.J., Spyker J.M. (1969). Beneficial effect of immobilization of joints in rheumatoid and related arthritides: a split study using sequential analysis. *Arthritis Rheum.*, **12**, 34.
20. Bland J.H., Eddy W.M. (1968). Hemiplegia and rheumatoid hemiarthritis. *Arthritis Rheum.*, **11**, 72.
21. McCarty D.J., Phelps P., Pyenson J. (1966). Crystal-induced inflammation in canine joints: an experimental model of quantification of host response. *J. Exp. Med.*, **124**, 99.

22. Wolff J. (1892). *Das Gesetz der Transformation der Knochen*. Berlin: Hirschwald.
23. Johnson L.C. (1975). Kinetics of skeletal remodelling. In *Birth defects: standard organisation of the skeleton*. Vol II. National Foundation – March of Dimes, p. 66.
24. Bullough P.G. (1981). The geometry of diarthrodial joints, its physiologic maintenance and the possible significance of age related changes in geometry-to-load distribution and the development of osteoarthritis. *Clin. Orthop.*, **156**, 61.
25. Rubin C.T., Lanyon L.E. (1984). Regulation of bone formation by applied dynamic loads. *J. Bone Joint Surg.*, **66A**, 397.
26. Nilsson B.E., Anderson S.M., Hardrup T.V. (1978). Bone mineral content in ballet dancers and weight lifters. *Proceedings of the 4th International Conference on Bone Mineral Measurement*. University of Toronto, p. 81.
27. Jones M.G., Watt D.G.D. (1971). Muscular control of landing from unexpected falls in man. *J. Physiol.*, **219**, 729.
28. Watson R.C. (1973). Bone growth and physical activity in young males. In *International Conference on Bone Mineral Measurement* (Mazess R.B. ed.). US DHEW: NIH, p. 380.
29. Simmons D.J., Russell J.E., Winter F. (1981). Space flight and the non weight bearing bones of the rat skeleton (Cosmos 1129). *Trans. Orthop. Res. Soc.*, **4**, 65.
30. Smith M.C., Rambout P.C., Vogel J.M. (1977). Bone mineral measurements. Experiment M078. In *Biomedical results from Skylab* (Johnston R.S., Diellein L.F. eds.). NASA publication.
31. Watson R.C. (1973). Bone growth and physical activity in young males. In *International Conference on Bone Mineral Measurement* (Mazess R.B. ed.). US DHEW: NIH, p. 352.
32. Szivek J.A., Cameron H.U., Weatherly G.C. (1981). A study of bone remodelling using biologically attached composite on-lay plates. *Trans. Orthop. Res. Soc.*, **6**, 61.
33. Crowninshield R.D., Brand R.H., Johnston R.C. (1980). Analysis of femoral component stem design in total hip arthroplasty. *J. Bone Joint Surg.*, **62A**, 68.
34. Chamay A., Tschantz P. (1972). Mechanical influence on bone remodelling. Experimental research on Wolff's law. *J. Biomech.*, **15**, 173.
35. Lanyon L.E., Goodship A.E., Pye A. (1982). Mechanically adaptive bone remodelling. A quantitative study of functional adaptation in the radius following ulnar osteotomy in sheep. *J. Biomech.*, **15**, 141.
36. Hert J., Liskova M., Landa J.L. (1971). Reaction of bone to mechanical stimuli: I. Continuous and intermittent loading of the tibia in the rabbit. *Folia Morphol. (Praha)*, **19**, 378.
37. Jaworski Z.F.G., Liskova-Kiar M., Uhthoff H.K. (1980). Effect of long term immobilisation in the pattern of bone loss in older dogs. *J. Bone Joint Surg.*, **62B**, 104.
38. Uhthoff H.K., Jaworski Z.F.G. (1978). Bone loss in response to long term immobilisation. *J. Bone Joint Surg.*, **60B**, 420.
39. Akeson S.H., Woo S.L.-Y., Amiel D. (1984). The chemical basis of tissue repair, the biology of ligaments. In *Rehabilitation of the Injured Knee* (Hunter L.Y., Funk F.J. eds.). St Louis: CV Mosby.
40. Roberts J., Burch T.A. (1966). Osteoarthritis prevalence in adults by age, sex, race and geographic area. *US Public Health Service Publication 1000*, Series 11, No. 15. Washington, D.C.
41. Meachim G. (1969). Age changes in articular cartilage. *Clin. Orthop.*, **64**, 33.
42. Sokoloff L. (1974). Cell biology and the repair of articular cartilage. *J. Rheumatol.*, **1**, 9.
43. Ah S.Y., Bayliss M.T. (1981). Cathepsin D and other proteases in human articular cartilage. *Semin. Arthritis Rheum.*, **XI**, 56.

44. Jasin H.E., Dingle J.T. (1981). Human Mononuclear cell factors mediate cartilage matrix degradation through chondrocyte activation. *J. Clin. Invest.*, **68**, 571.
45. Kellgren J.H., Lawrence J.S. (1958). Osteoarthritis and disk degeneration in an urban population. *Ann. Rheum. Dis.*, **17**, 388.
46. Lawrence J.S., Molyneux M. (1968). Degenerative joint disease amongst populations in Wensleydale (England) and Jamaica. *Int. J. Biometeorol.*, **12**, 163.
47. Department of Health, Education and Welfare (1979). Basic data on arthritis of the knee, hip and sacroiliac joint in adults aged 24–75 years. United States, 1971–1975. *NCHS* Series 14 No. 213: 79–1661. SHEW publications.
48. Forman M.D., Malamet R., Kaplan O. (1983). A survey of osteoarthritis of the knee in the elderly. *J. Rheumatol.*, **10**, 282.
49. Fox H. (1939). Chronic arthritis in wild mammals. *Trans. Am. Philos.*, **31**, 73.
50. Danielson L.G. (1964). Incidence and prognosis of coxarthrosis. *Acta Orthop. Scand.*, Supplement 66.
51. Seifert M., Whiteside C.G., Savage O. (1969). A five year followup of 50 cases of idiopathic osteoarthritis of the hip. *Ann. Rheum. Dis.*, **28**, 325.
52. Storey G.O., Landells J.W. (1971). Restoration of the femoral head after collapse in osteoarthrosis. *Ann. Rheum. Dis.*, **30**, 406.
53. Harrison M.H.M., Schajowicz F., Trueta J. (1953). Osteoarthritis of the hip. A study of the nature and evolution of the disease. *J. Bone Joint Surg.*, **35B**, 598.
54. Salter R.B., Simmonds D.F., Malcolm B.W., *et al.* (1980). The biological effect of continuous passive motion in the healing of full thickness defects in articular cartilage. An experimental investigation in the rabbit. *J. Bone Joint Surg.*, **62A**, 1232.
55. Ely L.W., Mensor M.C. (1933). Studies in the immobilisation of normal joints. *Surg. Gynaecol. Obstet.*, **57**, 212.
56. Salter P.B., Field P. (1960). The effects of continuous compression on living articular cartilage. *J. Bone Joint Surg.*, **42A**, 31.
57. Enneking W.F., Horowitz M. (1972). The intra-articular effects of immobilisation of the human knee. *J. Bone Joint Surg.*, **54A**, 973.
58. Crellin E.S., Southwick W.O. (1964). Changes induced by sustained pressure in the knee joint articular cartilage of adult rabbits. *Anat. Rec.*, **149**, 113.
59. Palmoski M., Perricone E., Brandt K.D. (1979). Development and reversal of proteoglycan aggregate defect in normal knee cartilage after immobilisation. *Arthritis Rheum.*, **22**, 508.
60. Sidney K.J., Shepherd R.J. (1976). Attitudes towards health and physical activity in the elderly. Effects of a physical training programme. *Med. Sci. Sports and Exerc.*, **8**, 246.
61. Shepherd R.J., Kavanaugh T. (1978). The effects of training on the aging process. *Physician Sports Med.*, **6**, 33.
62. Masoro R.J., Yu B.P., Bertrans H.A., *et al.* (1980) Nutritional problems of the aging process. *Federal Proc.*, **39**, 3178.
63. Munro H.N. (1982). Nutritional requirements in the elderly. *Hosp. Pract.*, **17**, 143.

Chapter 5

Preventing and Treating Stiff Joints

ROB HERBERT

INTRODUCTION

When joints are immobilized for long periods of time they may become stiff. This can be a problem, because the development of stiff joints can impede the performance of everyday motor tasks. In this chapter I will consider how physiotherapists might go about preventing the development of stiff joints, and how they can restore normal joint mobility to joints that have already become stiff. The chapter begins by clarifying what is meant by the terms *joint stiffness* and *stiff joints*, and this is followed by a description of some of the morphological changes that accompany the development of stiff joints. The remainder of the chapter will be spent constructing a rationale for therapeutic intervention and providing a brief review of what is known about the effectiveness of some commonly-used methods for preventing and treating stiff joints.

NORMAL JOINT STIFFNESS

The *stiffness* of a joint is one of its most important mechanical properties. It is important because it determines the amount of movement that results when torques, such as those produced by muscles or by gravity, act about the joint. A stiff joint will experience very little movement in response to a given torque, whereas a compliant joint, one that is not stiff, will experience a relatively large amount of movement in response to the same torque.

Joint stiffness is determined by the physical properties of the soft tissues which cross the joint. As the joint is moved some tissues are lengthened, and they resist lengthening in the same way as a spring does. That is, they develop tension that resists further lengthening and further joint movement. The degree to which a tissue resists lengthening can be illustrated with a length–tension curve[1–4] (Figure 5.1). Short, inextensible tissues produce large tensions at short lengths and therefore (other things being equal) they restrict joint movement more strongly than long and extensible tissues. With few exceptions, very little is known about which soft tissues provide the dominant resistance to movement in most human joints. It is probable

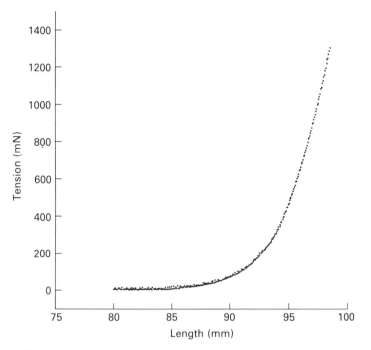

Figure 5.1   The length–tension curve of a whole rabbit soleus muscle–tendon unit.

though, that the major restraint to the so-called 'physiological' movements of most peripheral joints is provided by periarticular connective tissues (ligament and capsule), muscle, and, perhaps, skin.

A measure of a joint's stiffness can be derived from its torque–angle characteristics. (In the case of the so-called 'accessory movements',[5] which involve primarily translation rather than rotation of body segments, the stiffness of a joint can be derived from its force–displacement characteristics.[6–8]) To determine the torque–angle characteristics, the joint of a relaxed subject is moved to a known angle, and the externally applied torque required to maintain that joint angle is measured. If this is repeated at several different angles, a torque–angle curve can be constructed.[9–19] Part of the torque–angle curve of a human ankle joint has been reproduced in Figure 5.2. The stiffness of the joint is the rate of change of torque with respect to changes in joint angle. Graphically, the stiffness is found by determining the slope of the torque–angle curve. It can be seen from Figure 5.2 that the slope of the curve is small near the middle of the joint's range and large at the extremes of joint motion. That is, the joint is compliant near mid-range, but it becomes progressively stiffer as it is taken toward the end of range.

The definition of joint stiffness used in this chapter (i.e. the rate of change of passive joint torque with respect to change in joint angle) is consistent with the way in which stiffness is defined in mechanics, but it differs from the way in which the words are sometimes used by physiotherapists. With

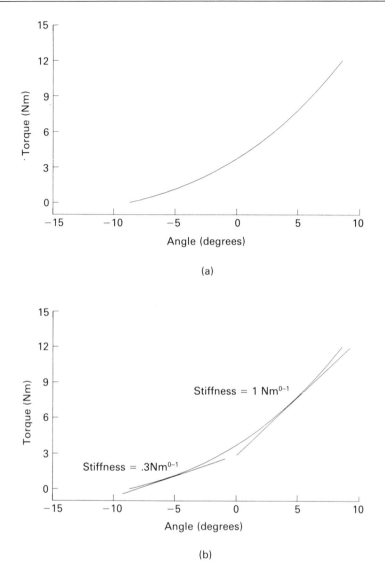

*Figure 5.2   (a) Part of the torque–angle relationship of a human ankle joint. Increasing joint angles indicate increasing dorsiflexion. (b) The stiffness of the joint is given by the slope of the torque–angle curve. As the joint is taken further into dorsiflexion (i.e. as larger torques are applied to the joint) the stiffness of the joint increases. (Adapted with permission from Chesworth B.M., Vandervoort A.A. (1988). Physiother. Can., **40**, 300.)*

the definition of joint stiffness used here, it is clear that all joints have a stiffness, because all joints demonstrate some resistance to movement. Also, with this definition, it is clear that any of the soft tissues that cross a joint can potentially contribute to its stiffness. In contrast, physiotherapists sometimes use the words joint stiffness to refer to increases in the stiffness of joints that are brought about by changes in the extensibility of periarticular connective tissues. In this chapter, joints with adaptively or pathologically increased stiffness will be referred to as *stiff joints*. Unless explicitly stated, the term stiff joints will be used regardless of whether the joint stiffness is predominantly due to periarticular connective tissues or to any other tissues that cross the joint.

STIFF JOINTS

Joints can become stiff for many reasons. Sometimes they become stiff as a result of the direct action of diseases on the soft tissues.[20] More commonly however, stiff joints develop, not as the direct result of disease, but in response to a change in the mechanical environment of soft tissues. If the soft tissues are deprived of movement and forces that they normally experience, the joint may become stiff.

The soft tissues may become deprived of movement and force in a variety of clinical circumstances. The deprivation of movement and force occurs, for example, when joints are immobilized by plaster casts or other fixation devices following injury, or when muscles are unable to generate sufficient active torque to move the joint through its normal range of movement. Also, people will sometimes voluntarily restrict joint movement at a painful joint because movement exacerbates their pain.

Clinical observations suggest that several factors determine the extent to which a joint's stiffness increases when it is deprived of movement. Perhaps not surprisingly, it appears that prolonged and complete immobilization can cause greater increases in joint stiffness than brief periods of partial immobilization. Also, when joints are immobilized following trauma they are particularly likely to become stiff, especially if the trauma was extensive or if it was associated with infection or marked oedema. One other factor that affects the development of joint stiffness is age. A pre-adolescent child's joints can sometimes be immobilized for long periods of time with apparently little effect, whereas an adult's joints, and particularly the joints of an elderly adult, may become very stiff when immobilized for the same period of time.

Observations made both on humans and animals provide some insights into the changes in tissue structure and composition that accompany the development of stiff joints. Some of these changes will be briefly described in the following section.

Muscle shortening

Physiotherapists have long recognized that immobilized muscles can become short enough to severely restrict joint range of motion. In fact, it is an easy matter to demonstrate muscle shortening in humans using objective clinical measures of the length and extensibility of multi-articular muscles.

However, there have been no studies that have systematically examined muscle-length changes in humans. As a result, most of what we know about how muscles adapt their length in response to immobilization comes from studies on animals.[21–23]

Animal studies clearly show that the adaptations of muscle length and extensibility that occur in response to immobilization depend upon the position in which the muscle is immobilized. When a muscle is immobilized in a shortened position (e.g. if the soleus muscle is immobilized by fixing the ankle in plantarflexion), or if a muscle is made to work predominantly in a shortened range, the passive length–tension curve of the muscle belly shifts to the left, and it may also become steeper.[24–27] That is, the muscle belly becomes shorter and stiffer. On the other hand, it has generally been found that when a muscle is immobilized in a lengthened position the muscle belly does not become short and the muscle belly passive length–tension characteristics remain unchanged.[24,25]

A recent study has investigated the effect of various positions of immobilization on the rest length of muscle–tendon units (i.e. of the muscle belly and its tendons) using rabbits whose ankle joints had been immobilized in plaster casts for 10 days.[28] As was expected on the basis of previous studies on muscle length adaptations, it was found that the muscle–tendon units became short when immobilized in a shortened position. The decreases in muscle–tendon unit rest length were approximately linearly related to position of immobilization (Figure 5.3a). In this study, only those muscle–tendon units immobilized in the most lengthened positions did not become short. Also, this study found no evidence of a position-dependent increase in muscle stiffness (Figure 5.3b), although there was evidence of an immobilization-induced increase in stiffness that was not dependent on position of immobilization.

The decreases in muscle length and extensibility observed following immobilization in a shortened position are accompanied by some remarkable changes in muscle morphology. Most strikingly, when muscles are immobilized in a shortened position they lose sarcomeres in series from their muscle fibres.[24–26] That is, contractile proteins are broken down and removed from the ends of the muscle. The magnitude of this loss can be very substantial; for example, a cat soleus muscle that has been immobilized in a shortened position may lose nearly 40% of its sarcomeres in series within 4 weeks.[24] If the period of immobilization is sufficiently long, the number of sarcomeres lost approximates that which brings about a shift in the muscle's optimal length (i.e. the length at which the muscle develops its greatest active tension) to the length at which the muscle was immobilized.[25] This implies that some mechanism exists to regulate the number of sarcomeres, and hence the length at which the muscle generates its greatest active tension, to the length at which the muscle is habitually used.

There have also been reports of immobilization-induced changes in the amount and alignment of the connective tissues in the muscle. In fact, while the relative amount of connective tissue in the muscle belly (i.e. the ratio of

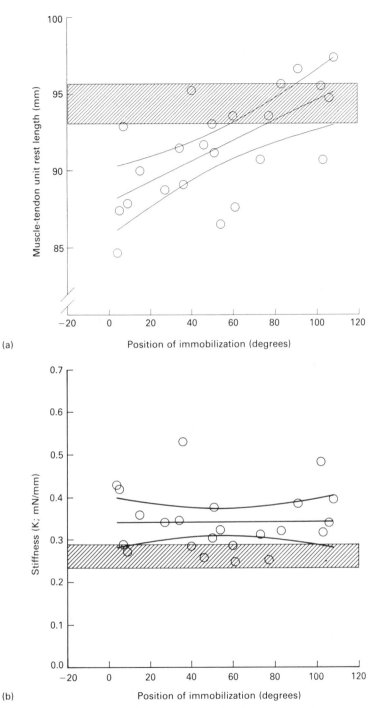

Figure 5.3   The effects of immobilization on muscle–tendon unit rest length and extensibility. (a) The effect of position of immobilization on muscle–tendon unit rest length. Each data point indicates the position of immobilization and post-immobilization length of a rabbit soleus muscle–tendon unit that has been immobilized in a plaster cast for ten days. Small joint angles correspond to immobilization in a shortened position. (b) The effect of position of immobilization on 'K', a measure of muscle–tendon unit stiffness. In both graphs the shaded area represents the 95% confidence interval about the mean for a group of muscles from non-immobilized rabbits.

connective tissue cross-sectional area to muscle fibre cross-sectional area) increases in response to immobilization in a shortened position,[25,29–32] it appears that the absolute amount of connective tissue actually *decreases* after the first few days of immobilization.[33,34] This indicates that immobilized muscles do not become 'fibrosed', as is sometimes thought. One report also suggests that the intramuscular connective tissue undergoes a change in its architecture when the muscle is immobilized in the shortened position.[30] In a normal muscle the intramuscular connective tissue is more or less transversely orientated when the muscle is not stretched,[30,35,36] but in adaptively shortened muscles it appears to become more longitudinally orientated at the same length.[30]

It is difficult to know which, if any, of these morphological changes is responsible for the decrease in muscle length and extensibility that is seen following immobilization in a shortened position. We can only reason that, because it is probable that both the intramuscular connective tissue and the intracellular structures associated with each sarcomere determine the length and stiffness of normal muscles,[36–43] it is possible that either changes in the intramuscular connective tissues or loss of sarcomeres could be responsible for adaptive shortening of muscle. Some very recent evidence suggests that architectural changes which accompany changes in fibre (and perhaps also tendon) morphology may also play an important role in producing immobilization-induced length changes in muscle.[44–46]

Decreases in the extensibility of periarticular connective tissues

At least in some circumstances, periarticular connective tissues also contribute to increased joint stiffness following immobilization. Some evidence for this has come from a series of studies on immobilized animal joints.[47–52] In these studies, measurements were made of the stiffness of immobilized joints after the skin and muscle had been removed from around the joints. Moderate durations of immobilization (e.g. 32 days of immobilization of monkey knees) were found to be associated with substantial increases in joint stiffness (Figure 5.4). This suggests that changes to tissues within or immediately around the joint (as opposed to muscle or skin) can be at least partly responsible for the increases in joint stiffness that develop in response to immobilization. In these studies, however, increases in the stiffness of the joints with muscle and skin removed was much smaller than the increases in stiffness observed when the muscle and skin were not removed.[48,53] This indicates that changes to intra-articular and periarticular connective tissues are less important causes of joint stiffness than changes in muscle and skin, at least with the experimental models of stiff joints used in these studies. It is not yet known how much changes in the extensibility of periarticular connective tissues contribute to the increases in joint stiffness that are observed in immobilized human joints.

Akeson and his colleagues have proposed a mechanism whereby immobilization could induce decreases in periarticular connective tissue extensibility. These investigators have observed that, in immobilized rabbit knees, the increase in joint stiffness is associated with changes in the biochemical composition of periarticular connective tissues.[47–49,51,54–56] The concentration of some glycosaminoglycans (large proteins in the connective tissue ground substance) decreases following prolonged immobilization, and this produces a substantial loss of water from the tissues. Normally the water in the tissues acts as a spacer that keeps collagen fibres apart. It has been hypothesized that the loss of water allows collagen fibres to come sufficiently close together to allow the formation of additional interfibre cross-links, and that the development of these cross-links causes some of the connective tissues to become inextensible and the joints to become stiff.[47–49,54–56]

While the work of Akeson and his colleagues has provided an important impetus to research on adaptations of periarticular connective tissue extensibility, some questions of fundamental importance are currently unanswered. Perhaps most significantly, it is not yet clear whether the

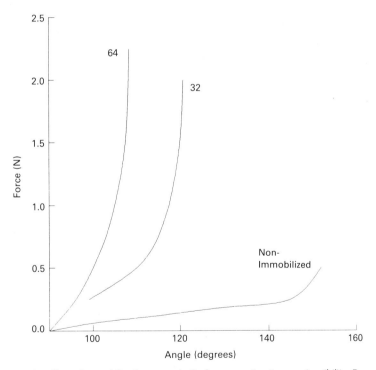

Figure 5.4   The effect of immobilization on periarticular connective tissue extensibility. Force–angle curves are of monkey knee joints with skin and muscle removed. (Force–angle curves are similar to torque–angle curves, except they relate joint angle to an applied force rather than to an applied torque.) The figure shows force–angle curves of a non-immobilized joint, and of joints immobilized for 32 and 64 days. With immobilization the joints become stiff, indicating that the periarticular connective tissues have become inextensible. (Adapted with permission from Lavigne A.B., Watkins R.P. (1973). In *Perspectives in Biomedical Engineering* (Kenedi R.M. ed.). London: Macmillan.)

adaptations of periarticular connective tissue extensibility exhibit the same dependence on position of immobilization as muscle. That is, it is not yet clear whether periarticular connective tissues become inextensible in response to immobilization regardless of position of immobilization, or only to immobilization in a shortened position. Some clinical observations would suggest that adaptations of periarticular connective tissue extensibility are position dependent. For example, it has been widely reported that 'shortening' of the collateral ligaments of the metacarpophalangeal joints occurs when these joints are immobilized in extension (i.e. when the ligaments are immobilized in a shortened position), but not when the joints are immobilized in flexion.[57,58]

Another important question is that of exactly which of the periarticular connective tissues cause the joints to become stiff. Studies on animals show clearly that immobilization brings about a *decrease* in stiffness of some ligaments.[59–63] This means that these ligaments at least cannot be responsible for the increases in joint stiffness that accompany immobilization. Clearly further research is needed to determine both how individual periarticular connective tissues respond to immobilization, and how they respond to immobilization in different positions.

Intra-articular adhesions

A third cause of joint stiffness is the development of intra-articular adhesions.[53,55,64,65] Several studies, on both animals and humans, have shown that with prolonged immobilization (several months in some animals or more than a year in humans) the joint space can become filled with a fibro-fatty tissue. With continued immobilization the fibro-fatty tissue becomes more fibrous, and it may eventually form strong bands of dense connective tissue that adhere to adjacent joint surfaces, preventing movement at the joint. An illustration of the strength of these adhesions is provided by one study in which it was observed that, when stiff joints were forcefully manipulated through a large range of joint angles, the cartilage to which adhesions were attached was torn from the joint surface, but the adhesions themselves remained intact.[64] This observation would seem to indicate that the development of intra-articular adhesions can be an important cause of stiff joints, at least in joints that have been immobilized for long periods of time.

Scar tissue

As we have already seen, joints are often immobilized following injury to soft tissues or bones. One cause of joint stiffness, the development of scar tissue, arises only following tissue injury, and it arises as an end product of the tissue repair process.[63,66–69] After the initial inflammatory response to injury, fibroblasts from nearby connective tissues migrate to the injured tissues and begin to produce and secrete a precursor of collagen. The subsequent production of collagen results in the formation of a fibrous scar. To a degree the scar can take on the characteristics of adjacent tissues; the scar that forms adjacent to bone or dense connective tissue is likely to be dense and inextensible, and the scar that develops adjacent to loose connective tissue is likely to be loose and extensible.[57,69] However the extent and extensibility of the scar is also determined by the severity of the initial inflammatory response. Severe injury with extensive oedema is therefore likely to produce extensive, dense scarring.[57,69–71]

It would appear that the formation of scar tissue could bring about an increase in joint stiffness in a number of ways. For example, a previously compliant tissue that has been repaired with dense scar may become inextensible, in which case it might strongly resist lengthening, and thereby resist joint motion. Perhaps more importantly, scar can tether adjacent tissues to each other, preventing the relative movement of tissues that normally accompanies joint motion. The development of scar is therefore a particular problem in the hand, where there are a large number of structures moving relative to one another and through narrow spaces such as the fibro-osseus tunnels. Most disabling is the formation of scar, or 'adhesions', which tether tendons to bone, ligaments or skin. These adhesions are a notorious cause of stiff joints following flexor tendon surgery.[57,70,72–74] It was once thought that tendon adhesions were a necessary consequence of tendon surgery because they represented an extrinsic source of repair to a tissue that was otherwise incapable of repair. However, it is now thought that, at least under experimental conditions, tendon is capable of intrinsic repair, and that the development of adhesions is not necessary for tendon repair to proceed.[73–78]

A RATIONALE FOR THERAPEUTIC INTERVENTION

When a joint becomes stiff its torque–angle curve, or part of its torque–angle curve, becomes steeper (Figure 5.5).[79] Consequently, when forces or torques such as those produced by actively contracting muscles act to move the joint, less movement results. This restriction of joint movement can cause profound movement dysfunction. For example, a person whose ankle is stiff in dorsiflexion may be unable to dorsiflex adequately when standing up from a seated position. The lack of dorsiflexion may interfere with efficient standing up, or, if the movement control system is unable to adequately compensate, it may prevent the person from standing up at all. Physiotherapists often talk of a 'decreased range of joint motion', because it is decreased joint motion or mobility that is the cause of the movement dysfunction. Physiotherapists are concerned with the prevention and treatment of stiff joints primarily because decreased joint motion can prevent the performance of everyday motor tasks.

The cornerstone of therapeutic intervention aimed at preventing and treating stiff joints is the provision of movement and stretch. There are several theories cited in the literature of how movement or stretch can prevent or reverse excessive joint stiffness. One theory says that stretching of soft tissues can tear adhesions that limit joint movement, and in this way stretching can restore normal joint mobility. The tearing of adhesions has been said to cause the increases in joint range of motion that are seen immediately following the manipulation of stiff joints.[80,81] Also, it is probable that manipulation under anaesthesia, a procedure sometimes used by surgeons to restore normal motion to chronically stiff joints, increases joint range of motion by tearing intra-articular adhesions or other soft tissues.[81] Generally, however, the tearing of tissues is not considered by most physiotherapists to be the mechanism by which physiotherapy brings about increases in joint compliance or joint range of motion. In fact, forceful stretching techniques have often been discouraged,[57,71,72] especially when treating joints that have become stiff following injury and subsequent

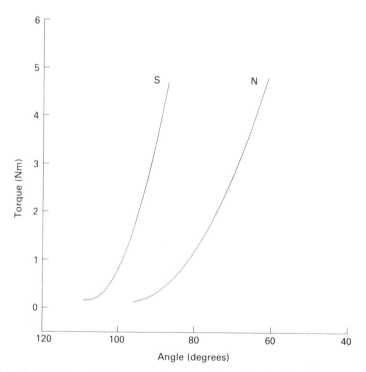

Figure 5.5 A comparison of the torque–angle curves of normal and stiff joints. Decreasing angles indicate increasing degrees of dorsiflexion. Curve N is the torque–angle curve for the ankle of a normal 6-year-old child. Curve S is the torque–angle curve of a 6-year-old child with cerebral palsy. A 5 Nm torque produces a displacement of the stiff ankle joint that is about 25° less than the displacement produced by the same torque on the normal ankle joint. (Adapted with permission from Tardieu C., Huet de la Tour E., Bret M.D. et al. (1982) Dev. Med. Child Neurol., 63, 97.)

immobilization, because it is said that forceful stretching can cause tearing of tissues, which promotes further scar formation and increases joint stiffness.

Another theory, which appears to be the dominant one in physiotherapy literature, is based on observations of the time-dependent mechanical behaviour of joints or soft tissues in response to sustained stretch. It has been widely documented that the length of soft tissues, and hence the range of motion available at joints, can be increased with the provision of sustained stretching. Immediately after a period of stretching soft tissues exhibit less resistance to lengthening than they did before they were stretched,[1–4,82–87] and consequently, after stretching, the joint undergoes greater displacement at any given applied torque[12,88] (Figure 5.6). The magnitude of the increase in displacement (or increase in range of motion) is dependent on the duration of the applied stretch – if the stretch is applied for long periods, the increase in displacement is large – hence the term 'time-dependent' deformation. There is a widely held belief that the progressive increase in joint range of motion which is observed when stiff joints are stretched is brought about by an accumulation of the time-dependent deformation that occurs with successive stretches. Proponents of this theory hold that, because time-dependent deformation is greatest with long durations of stretch, sustained stretching is more effective than brief stretching for the treatment of stiff joints.[87,89–96] Also, because time-dependent deformation is temperature-dependent, this theory has been used to justify the use of 'heat and stretch' procedures in which the tissues are heated before stretching. It is said that because heating the tissues increases time-dependent deformation, continued heat and stretch will produce greater increases in joint range of motion than sustained stretching alone.[91,92]

While there is little doubt that time-dependent deformation is responsible for the increase in joint range of motion seen immediately after stretching, it is unlikely that time-dependent deformation could bring about a lasting restoration of normal joint mobility to stiff joints. This is because the time-dependent deformation produced by forces smaller than those that tear tissues is largely a transient phenomena. Over the period of time that follows the stretch (perhaps a few hours) the joint's range of motion will return towards pre-stretch values.[95] Moreover, the tearing and time-dependent deformation of tissues are unlikely, on their own, to bring about a lasting restoration of the normal mechanical properties of joints, because neither the tearing of tissues nor their time-dependent deformation involves a restoration of normal tissue structure or composition. Time-dependent deformation of connective tissues, for example, is the result of collagen fibres being made to slide past each other in the viscous ground substance in which they are embedded, and the tearing of connective tissues occurs when the collagen fibres become stretched to the point at which they rupture. These processes are purely physical, i.e. they are simply the result of mechanical deformation of the tissues, and they do not require the action of biological processes such as protein synthesis and lysis which could

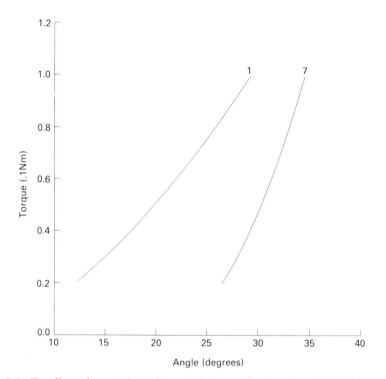

Figure 5.6   The effects of repeated stretching on joint range of motion. Curve 1 is the torque–angle curve of an MCP joint complex in the direction of abduction. Curve 7 is the torque–angle curve of the same joints after six 75-second stretches have been applied to the joint. After stretching, the range of motion that corresponds to a 0.75 Nm torque has increased by about 8°. This is an example of time-dependent deformation of soft tissues. Adapted with permission from Loebl W.Y. (1972). (Rheumatol. Phys. Med., **8**, 365.)

bring about a change in soft tissue structure and composition. Yet as we have seen, the soft tissues of stiff joints differ, both in their composition and structure, from the soft tissues of joints with normal stiffness. Presumably therefore, the mechanism by which normal mobility is restored to stiff joints must involve biological processes capable of changing the composition and structure of the soft tissues, and not just mechanical deformation. The time-dependent increases in joint range of motion observed immediately after stretching may therefore be unrelated, both in mechanism and magnitude, to the lasting changes in joint mobility produced by adaptations of tissue composition and structure.

If, as has been argued above, the effective prevention and treatment of stiff joints involves the prevention of changes in, or the restoration of, normal tissue composition and structure, then we must consider how it is that physiotherapists can influence tissue composition and structure. The answer lies in the observation that, as we have already seen, soft tissue structure and composition are regulated in part by the mechanical environment of the tissues. That is, the tissues adapt their structure and composition, and therefore their length and extensibility, in response to the forces or movement that they experience. It follows that physiotherapists can control adaptations of tissue structure and composition, and therefore the stiffness of the joint, by manipulating the mechanical environment of the soft tissues.

How should physiotherapists go about manipulating the mechanical environment of joints, and what forces and movements should they provide to prevent an immobilized joint becoming stiff, or to restore normal joint mobility to an already stiff joint? The simplest answer is that if the soft tissues can be made to experience forces and movements that are similar to the forces and movements experienced during the performance of everyday activity, then it is likely that a sufficient stimulus will be provided for the soft tissues to adapt their structure and composition so that normal joint stiffness is maintained or restored. We know that the performance of everyday movement provides the stimulus for maintenance of normal joint stiffness, at least in the absence of soft tissue trauma, because generally joints that experience normal motor activity do not become stiff, and joints that are deprived of normal movement and forces do. Also, observations on animals have shown that, providing the period of immobilization is not too prolonged, unrestrained activity following immobilization can provide a sufficient stimulus to reverse post-immobilization restrictions of joint range of motion.[49,53,65]

In many circumstances people may become unable to move their joints in a way that causes the soft tissues to experience the movement and forces required to prevent the development of stiff joints. Also, once people have already developed stiff joints they may become unable to move in a way that will provide the most effective stimulus for adaptations that will restore normal joint stiffness. There is a widespread belief amongst physiotherapists that, in these circumstances, stiff joints can be prevented and normal joint stiffness can be most quickly restored if the physiotherapist can ensure that the soft tissues experience some critical level of movement or stretch. It is this assumption that underlies the use of therapeutic movement and stretch aimed at preventing and treating stiff joints.

Typically physiotherapists provide movement or stretch to soft tissues using one or a few of a number of treatment techniques, such as passive movements, passive joint mobilizations, manual stretching, prolonged passive stretching, continuous passive motion, splinting and serial casting. With treatment techniques such as these, physiotherapists try to provide the mechanical environment required for the maintenance of normal joint compliance outside of the usual context of everyday motor activity. In using these techniques the assumption is implicitly made that not all of the characteristics of the complex mechanical environment provided by everyday motor activity are of equal importance in regulating joint compliance. That is, the assumption is made that some features of the mechanical environment are particularly important in regulating joint stiffness. In this chapter those parameters will be referred to collectively as the *critical mechanical stimulus*. In effect, physiotherapists make assumptions about the nature of the critical mechanical stimulus, and they apply what is assumed to be that stimulus outside of the context of the performance of motor tasks, in order to prevent the development of stiff joints or to restore normal joint stiffness.

Of course if the development of stiff joints is to be permanently restored, the person must eventually become able to perform everyday motor tasks in a way that provides the necessary mechanical stimulus for the soft tissues. Until this occurs, the withdrawal of therapy will inevitably be accompanied by the development of stiff joints. In this respect the motor learning model of rehabilitation, proposed by Carr and Shepherd,[97,98] provides a useful framework within which to view these widely-used treatments.

In the motor learning model, the problem of stiff joints is not seen to be independent of the broader problem of an inability to perform motor tasks. The model emphasizes that there is little point in physiotherapists providing the critical mechanical stimulus to the soft tissues if the person with the stiff joint does not ultimately learn to move in a way that causes the tissues to experience the critical mechanical stimulus without the intervention of a physiotherapist. In many people with musculoskeletal problems of reasonably short duration, a lasting inability to move in a way that provides the critical mechanical stimulus to soft tissues is unlikely to be a problem. For example, following a short period of immobilization associated with a minor knee injury most people will return quite quickly to a pattern of walking that provides the stimulus to maintain normal joint stiffness. On the other hand, when a person with long-standing osteoarthritis of the hip undergoes hip arthroplasty they may continue to walk in a way that does not provide a sufficient stimulus to, say, maintain the length of the iliopsoas muscle, even though such a pattern of walking is no longer prevented by pain. In this case, the stretch or movement provided by traditional physiotherapy techniques is unlikely to produce a maintained restoration of normal joint stiffness, because when treatment ceases, the soft tissues will no longer experience the necessary mechanical stimulus. Effective therapy in this case must involve teaching the person better (presumably normal) movement strategies.

THE EFFECTIVENESS OF PHYSIOTHERAPY INTERVENTIONS

In the remainder of this chapter I will briefly review those studies which tell us about how immobilized joints and stiff joints adapt their stiffness in response to movement or stretch. Particular attention will be paid to the small number of studies which have investigated the effectiveness of clinical techniques used to prevent the development of stiff joints, or to restore normal joint stiffness to already stiff joints.

Preventing the development of stiff joints

In the preceding section I have argued that when physiotherapy techniques bring about lasting changes in joint stiffness they do so by stimulating adaptations of soft tissue composition and structure. The intent of physiotherapy intervention should therefore be to provide an appropriate and sufficient mechanical stimulus for the maintenance of normal tissue structure and composition. Sometimes, as when the joint is immobilized by a rigid cast, severe pain or acute injury, it is not possible to provide movement and stretch to the immobilized soft tissues. In this case, physiotherapists can do little to prevent the development of stiff joints, except perhaps to minimize the development of oedema where this is likely to occur. However, where it is possible to provide movement or stretch to soft tissues, physiotherapists have traditionally tried to prevent the development of stiff joints by providing small numbers of cyclical movements to the joint, usually in the form of passive or active movements, or joint mobilizations. Remarkably, while there are numerous anecdotal reports of the effectiveness of these treatments, there have been very few experimental studies that have investigated their efficacy.

A small number of animal studies provide mixed support of the effectiveness of cyclical movements at preventing the development of stiff joints. One of these studies investigated the effect of 5 minutes of daily 'resistance-free exercise' on the development of stiffness in otherwise immobilized rabbit knee joints.[99] Although this study did not use inferential statistics, the authors concluded that there was little effect of the exercises on the development of stiff joints. In fact they concluded that overly vigorous exercise may actually cause injury to soft tissues. Similar conclusions have been drawn by physiotherapists on the basis of clinical observations.[97] It has been suggested that the injudicious use of passive movements, particularly forced full range passive movements to shoulder joints, may cause soft tissue injury and the development of painful joints.

A series of studies, which have investigated the effectiveness of passive movements on the paws of dogs immobilized after flexor tendon repair, have had more positive results. These studies have shown that dogs that received 5 minutes daily of passive movements through part of their range of motion had significantly greater joint range of motion after 6 weeks of immobilization than dogs that were immobilized without receiving passive movements.[100,101] Dogs that received daily passive movements did not develop adhesions between the tendon and its sheath, whereas completely immobilized dogs did.[101,102] These results are encouraging because they suggest that relatively small numbers of passive movements can substantially influence the properties of tendon adhesions and can influence the development of joint stiffness.

Most of the studies that have investigated the effect of cyclical movements on the development of stiff joints in humans have been performed on people who have had surgical repairs of flexor tendons of the hand. These studies have generally concluded that when immobilization is interrupted with early active or passive movements sooner rather than later following surgical repair, the development of stiff joints can be retarded.[72,103–111] Unfortunately none of these studies have utilized strong experimental designs.[112,113] Usually, for example, these studies involve non-random allocation of subjects to early motion and control groups, or they use literature controls. For this reason, these studies do not provide strong evidence of the effectiveness of early controlled motion in preventing the development of stiff joints.

Some investigators have tried to optimize the effectiveness of cyclical movements by providing large numbers of movements to joints. This has become possible with the development of clinically feasible continuous passive motion (CPM) machines, which can be used to apply large numbers of cyclical movements to otherwise immobilized joints. CPM machines are now widely used in many clinical settings, particularly for preventing the development of joint stiffness following arthroplasty. It is thought that the ability of CPM to provide a large number of joint movements might make it a particularly effective way of preventing the development of stiff joints. The implicit assumption is that the number of movement cycles is an important parameter of the critical mechanical stimulus and that large numbers of movement cycles can more effectively prevent the development of stiff joints.

A number of clinical studies have investigated the efficacy of CPM[114–128] but the findings of these studies do not provide strong support of the effectiveness of CPM. Four clinical trials, all on people who had undergone knee arthroplasty, have employed truly random allocation of subjects to treatment and control groups.[114,124,125,127] The control groups received traditional physiotherapy stretches or exercise programmes. In these studies CPM was used for an average of between 5 and 9 days post-operatively, and for between 4 and almost 24 hours per day. Two of these studies demonstrated a significant effect of CPM (with a mean effect of between about 6 and 16° of range of motion), and the other two studies did not find a significant effect of CPM. Significant effects of CPM were only found in the early post-operative period (up to 17 days post-operatively); no effect was found at 2, 3, 6 or 12 months post-operatively.[125,127] It may be that the optimal way of providing CPM has yet to be found, but there is little evidence from clinical trials to suggest that CPM is likely to be substantially more effective than other less sophisticated procedures used to prevent the development of stiff joints.

A very different method used to prevent stiff joints is sustained stretch. Physiotherapists use sustained stretch to prevent the development of stiff joints when they position the limbs of people with paralysis or when they use splints or casts to hold soft tissues in a lengthened position. A common feature of these techniques is that the joint is taken toward one extreme of range, often until there is a substantial tension in the soft tissues, and it is held in that position for a period of time. The assumption is that the critical mechanical stimulus is related to the duration for which the soft tissues experience substantial tensions. Also, in selecting these techniques, physiotherapists implicitly assume that the number of times that the tissues are stretched is not an important parameter of the critical mechanical stimulus.

There is some evidence from animal studies that sustained stretch can prevent the development of stiff joints. For example, we have already seen that if muscles are immobilized in a lengthened position (i.e. if they receive a constant stretch) then they do not become as short as do muscles immobilized in a shortened position. Moreover, it has been shown that short, intermittent periods of stretch can prevent adaptive muscle shortening in otherwise immobilized mouse soleus muscles.[129–131] In these studies, mouse ankle joints were immobilized in a plantarflexed position, causing the soleus muscle to become short and inextensible. Some of the mice however, had their casts removed daily for varying periods of time, during which the ankle was held in a dorsiflexed position. It was found that 15 minutes of stretch each day (the shortest period of stretch studied) was sufficient to substantially retard the losses of range of motion that occurred with uninterrupted immobilization. One half hour of stretch daily was sufficient to completely prevent losses of joint range of motion. These results suggest that muscles do not adapt their length in response to the average length at which they are held during the day. Instead, it appears that muscles are particularly sensitive to the most extreme lengths that they experience, even though these extreme lengths might be attained for only relatively short durations. These animal studies would suggest that sustained stretch is likely to be an effective way of preventing muscle shortening and the development of stiff joints.

Evidence of the ability of soft tissues to adapt to sustained stretch has been used by some physiotherapists to develop clinical strategies for preventing the development of stiff joints. Some physiotherapists, for example, have advocated regular, strategic positioning of the limbs of unconscious or paralysed people.[97,132,133] It is thought that this may put sufficient tension on soft tissues for sufficient duration to prevent unwanted soft tissue adaptations. Also, physiotherapists have tried to keep soft tissues on stretch with the use of serial casting in order to prevent disabling increases in joint stiffness.[134–137] Three clinical studies have variously found a substantial effect,[134,135] or no effect,[136] of serial casting in preventing plantarflexor shortening in acutely head-injured people, although none of these studies used strong experimental designs. Therefore, while animal studies suggest that sustained stretch can prevent the development of adaptive muscle shortening, the few clinical studies thus far provide equivocal evidence of the effectiveness of sustained stretch for preventing the development of stiff joints.

Restoration of normal compliance to stiff joints

A quarter of a century ago Kottke and his colleagues popularized the use of sustained stretching (stretches, say, of 20 minutes duration or longer) for the treatment of stiff joints.[89] On the basis of their clinical observations, and with the use of the rationale that increases of range of motion are brought about by the accumulation of time-dependent deformation of soft tissues, they championed the superiority of sustained stretching over stretches of shorter duration. As we have seen, the rationale used by Kottke and his colleagues is dubious; it is unlikely that sustained increases in joint range of motion are brought about by time-dependent deformation of soft tissues. However, there is some evidence that sustained stretching can provide a powerful stimulus for the soft tissue adaptations required to restore normal stiffness to stiff joints.

The most important evidence for this comes from a study by Light *et al.*[138] This study is important because it is the first research that has used a true experimental design[112] to compare the effectiveness of physiotherapy treatments used to treat stiff joints. In this study, nursing-home residents with bilateral knee flexion contractures had both their right and left legs randomly allocated to either a high load, brief stretch group or a low load, prolonged stretch group. Legs in the high load, brief stretch group received ten passive movements and 1 minute of forced passive stretching into extension, applied manually by a physiotherapist, every day. Legs in the low load, prolonged stretch group were stretched into extension by applying traction to the leg, with weights of between 5 and 12 lb, for 1 hour each day. With only one exception, all legs that received the prolonged stretch experienced a greater increase in range of motion over the 4-week experimental period. The increases in range of motion of the prolonged stretch group were significantly greater than those of the brief stretch group. Interestingly, these results demonstrate that prolonged stretching was more effective than brief stretching, despite the fact that the sustained stretch group presumably experienced smaller loads (torques) than the brief stretch group. Two conclusions can be drawn from this study. Firstly, it demonstrates the superiority of sustained stretching over brief stretching techniques, at least in the treatment of knee flexion contractures. Secondly, if one accepts that brief stretch is unlikely to be *less* effective than no treatment, then this study provides convincing evidence that prolonged stretch is *more* effective than no treatment for increasing range of motion in stiff joints.

When physiotherapists use sustained stretch to restore mobility to already stiff joints, they typically stretch more aggressively than when they are stretching to prevent the development of stiff joints. This is reflected in the manner in which the stretch is commonly applied. Physiotherapists may use weights or serial casts or splints, including splints which utilize springs or turnbuckles, to progressively stretch the soft tissues which cross stiff joints. There is a widespread belief amongst physiotherapists, and a large number of published anecdotal and non-experimental reports, that these techniques represent a powerful way of restoring mobility to stiff joints[89,90,139–146] but, there have been few experimental investigations of the efficacy of these procedures. A recent study has conclusively demonstrated that serial casting can significantly reduce ankle stiffness in head-injured people with clinically evident plantarflexor muscle shortening.[147]

A different approach to the treatment of stiff joints is the use of cyclical or 'oscillatory' movements. Maitland[5,80,148] has advocated a system of joint assessment and treatment that utilizes repeated cyclical loading of soft tissues for the treatment of stiff joints (joint mobilization), and this system is widely used by physiotherapists. Only one study has experimentally investigated the effect of joint mobilization on stiff joints.[149] In this study, twelve dogs had their carpal joints immobilized in casts for a period of 6 weeks, following which they demonstrated substantially restricted range of carpal joint motion. For the ensuing 4 weeks, some dogs received daily joint mobilizations to their carpal joints, and others did not. All dogs were allowed to freely ambulate in their cages during this period. At the end of the experimental period the dogs that had received joint mobilizations showed statistically greater ranges of carpal joint range of motion than dogs that did not receive joint mobilizations, suggesting that joint mobilizations did indeed have a treatment effect. However, the mean range of motion difference between the two groups (an average of 2° difference across the 4-week treatment period) was small. If the findings of this study are taken at face value they could indicate that mobilizations have only a small effect on joint stiffness, or that mobilizations have little effect over and above the mobilizing effect of ambulation.

SUMMARY

The stiffness of a joint is important because it determines how much movement will result when a torque is applied to the joint. Joints can become stiff when they are deprived of the forces and movement that they normally experience. The development of stiff joints is a consequence of changes that occur in the soft tissues: muscles may become short when they are immobilized in a shortened position, and periarticular connective tissue can respond to immobilization by becoming inextensible. In addition, when joints are immobilized for long periods of time or if they are immobilized following trauma, they may develop fibrous adhesions that limit the movement between joint surfaces or between soft tissues. The increases in joint stiffness that result from soft tissue adaptations can be a problem because they can restrict the joint movement necessary for the performance of everyday motor tasks.

Prevention and treatment of stiff joints implies the prevention or reversal of the changes in soft tissue composition and structure that cause joints to become stiff. Effective therapy is, therefore, that which stimulates appropriate adaptations of soft tissue composition and structure. Commonly physiotherapists try to influence these adaptations by providing movement or stretch to the soft tissues. In doing so they make implicit assumptions about the nature of the critical mechanical stimulus that triggers soft tissue adaptations.

There has been very little research into the effectiveness of procedures that are used for preventing and treating stiff joints. Some experimental studies on animals, and some non-experimental studies on humans, suggest that quite small numbers of passive movements can prevent the development of tendon adhesions which cause joints to become stiff following tendon injury. It is not yet clear whether the large numbers of movements provided by CPM machines can provide a way of preventing the development of stiff joints that is superior to small numbers of passive movements performed manually by physiotherapists. There is some evidence from a small number of animal studies that quite short periods of sustained muscle stretching can prevent adaptive muscle shortening, although the few relevant human studies are inconclusive. Clinical trials on humans suggest that prolonged stretching is more effective than brief stretching in increasing the range of motion of stiff knee joints, and they indicate that serial casting is an effective means of lengthening short plantarflexor muscles in a head-injured population.

ACKNOWLEDGEMENTS

I would like to thank Roberta Shepherd, Janet Carr and Michael Lee for their helpful comments on an earlier draft of this chapter.

REFERENCES

1. Viidik A. (1966). Biomechanics and functional adaptation of tendons and joint ligaments. In *Studies of the Anatomy and Function of Bone and Joints* (Evans F.G. ed.). Berlin: Springer-Verlag.
2. Viidik A. (1973). Functional properties of connective tissues. *Int. Rev. Connect. Tissue Res.*, **6**, 218.
3. Viidik A. (1979). Biomechanical behaviour of soft connective tissues. In *Progress in Biomechanics* (Akkas N. ed.). Alphen aan den Rijn: Sijthoff and Noordhof.
4. Butler D.L., Grood E.S., Noyes F.R. (1979). Biomechanics of ligaments and tendons. *Exerc. Sports Sci. Rev.*, **6**, 125.
5. Maitland G.D. (1977). *Peripheral Manipulation*, 2nd edn. London: Butterworth.
6. Butler D.L., Noyes F.R., Grood E.S. (1980). Ligamentous restraints to anterior-posterior draw in the human knee. *J. Bone Joint Surg.*, **62A**, 259.
7. Tindle P. (1987). Force-displacement curves of the knee. *Proceedings of the 5th Biennial M.T.A.A. Conference*, Melbourne.

8. Lee M., Svensson N.L. (1990). Measurement of stiffness during simulated spinal physiotherapy. *Clin. Phys. Physiol. Meas.*, **11**, 201.

9. Chesworth B.M., Vandervoort A.A. (1988). Reliability of a torque motor system for measurement of passive ankle joint stiffness in control subjects. *Physiother. Can.*, **40**, 300.

10. Chesworth B.M., Vandervoort A.A. (1989). Age and passive ankle stiffness in healthy women. *Phys. Ther.*, **69**, 217.

11. Toft E., Espersen G.T., Kalund S. (1989). Passive tension of the ankle before and after stretching. *Am. J. Sports Med.*, **17**, 489.

12. Loebl W.Y. (1972). The assessment of mobility of metacarpophalangeal joints. *Rheumatol. Phys. Med.*, **8**, 365.

13. Breger-Lee D., Bell-Krostoski J., Brandsma J.W. (1990). Torque range of motion in the hand clinic. *J. Hand Ther.*, January–March.

14. Hufschmidt A., Mauritz K.-H. (1985). Chronic transformation of muscle in spasticity: a peripheral contribution to increased tone. *J. Neurol. Neurosurg. Psychiatry*, **48**, 676.

15. Engin A.E. (1985). Passive and active resistive force characteristics in major human joints. In *Biomechanics of Normal and Pathological Human Articulating Joints* (Berme N., Engin A.E., Corriea da Silva eds.). Dordrecht: Martinus Nijhoff, p. 137.

16. Mansour J.M., Audu M.L. (1986). The passive elastic moment at the knee and its influence on human gait. *J. Biomech.*, **19**, 369.

17. Unsworth A., Yung P., Haslock I. (1982). Measurement of stiffness in the metacarpophalangeal joint: the arthrograph. *Clin. Phys. Physiol. Meas.*, **3**, 273.

18. Weigner A.W., Watts R.L. (1986). Elastic properties of muscles measured at the elbow in man: I. Normal controls. *J. Neurol. Neurosurg. Psychiatry*, **49**, 1171.

19. Heerkens Y.F., Wottiez R.D., Huijing P.A. (1985). Inter-individual differences in the passive resistance of the human knee. *Human Movement Sci.*, **4**, 167.

20. Rodnan G.P., Schumacher H.R., eds. (1983). *Primer on the Rheumatic Diseases*. 8th edn. Atlanta: Arthritis Foundation.

21. Herbert R. (1988). The passive mechanical properties of muscle and their adaptations to altered patterns of use. *Aust. J. Physiother.*, **34**, 141.

22. Gossman M.R., Sahrmann S.A., Rose S.J. (1982). Review of length-associated changes in muscle. Experimental evidence and clinical implications. *Phys. Ther.*, **62**, 1799.

23. O'Dwyer N.J., Nielson P.D., Nash J. (1989). Mechanisms of muscle growth related to muscle contracture in cerebral palsy. *Dev. Med. Child Neurol.*, **31**, 543.

24. Tabary J.C., Tabary C., Tardieu C., *et al.* (1972). Physiological and structural changes in the cat's soleus muscle due to immobilisation at different lengths by plaster casts. *J. Physiol.*, **224**, 231.

25. Williams P.E., Goldspink G. (1978). Changes in sarcomere length and physiological properties in immobilised muscle. *J. Anat.*, **127**, 459.

26. Witzmann F.A., Kim D.H., Fitts R.H. (1982). Hindlimb immobilisation: length-tension and contractile properties of skeletal muscle. *J. Appl. Physiol.*, **53**, 335.

27. Supinski G.S., Kelsen S.G. (1982). Effect of elastase-induced emphysema on the force-generating ability of the diaphragm. *J. Clin. Invest.*, **70**, 978.

28. Herbert R.D., Balnave R.J. (1993). The effect of position of immobilisation on resting length, resting stiffness and weight of rabbit soleus muscle. *J. Orthop. Res.* In press.

29. Goldspink G., Williams P.E. (1978). The nature of the increased passive resistance in muscle following immobilisation of the mouse soleus muscle. *J. Physiol.*, **289**, 55P.

30. Williams P.E., Goldspink G. (1984). Connective tissue changes in immobilised muscle. *J. Anat.*, **138**, 343.

31. Williams P.E. (1989). Intramuscular connective tissue changes associated with different activity patterns. In *Motor Disturbances: Mechanisms and Implications for Therapy* (Torode M., Balnave R. eds.). Sydney: Cumberland College of Health Sciences.

32. Jocza L., Kannus P., Thoring J. (1990). The effect of tenotomy and immobilisation on intramuscular connective tissue. A morphometric and microscopic study in rat calf muscles. *J. Bone Joint Surg.*, **72B**, 293.

33. Williams P.E. (1981). Changes in the connective tissue component of muscle during periods of decreased or increased activity. *J. Anat.*, **133**, 133.

34. Williams P.E. (1989). Personal communication.

35. Rowe R.W.D. (1974). Collagen fibre arrangement in intramuscular connective tissue. Changes associated with muscle shortening and their possible relevance to raw meat toughness measurements. *J. Food Technol.*, **9**, 501.

36. Borg T.K., Caulfield J.B. (1980). Morphology of connective tissue in skeletal muscle. *Tissue Cell*, **12**, 197.

37. Magid A., Law J. (1985). Myofibrils bear most of the resting tension in frog skeletal muscle. *Science*, **230**, 1280.

38. Purslow P.P. (1989). Strain-induced reorientation of an intramuscular connective tissue network: implications for passive muscle elasticity. *J. Biomech.*, **22**, 21.

39. Hill A.V. (1952). The thermodynamics of elasticity in resting striated muscle. *Proc. R. Soc. Lond.*, **139**, 464.

40. Magid A., Ting-Beall H.P., Carvell M. (1984). Connecting filaments, core filaments and side-struts: a proposal to add three new load-bearing structures to the sliding filament model. In *Contractile Mechanisms in Muscle* (Pollack G.H., Sugi H. eds.). New York: Plenum Press.

41. Wang K. (1984). Cytoskeletal matrix in striated muscle: the role of titin, nebulin and intermediate filaments. In *Contractile Mechanisms in Muscle* (Pollack G.H., Sugi H. eds.). New York: Plenum Press.

42. Fish D., Orenstein J., Bloom S. (1984). Passive stiffness of isolated cardiac and skeletal myocytes in the hamster. *Circ. Res.*, **54**, 267.

43. Maruyama K., Matsubara S., Natori R. (1977). Connectin, an elastic protein of muscle. *J. Biochem.*, **82**, 317.

44. Huijing P.A., Heslinga J.W. (1991). Muscle fibre (hyper-) trophy and atrophy in relation to fibre angle. *Proceedings of the XIIIth International Congress on Biomechanics*, Perth, 14.

45. Heslinga J.W. (1992). The effects of immobilisation on muscle architecture and function in relation to normal growth. PhD thesis, Vrije University.

46. Halar E.M., Stolov W.C., Venkatesh B., *et al.* (1978). Gastrocnemius muscle belly and tendon length in stroke patients and able-bodied persons. *Arch. Phys. Med. Rehabil.*, **59**, 476.

47. Akeson W.H., Amiel D., Woo S.L.-Y. (1980). Immobility effects on synovial joints. The pathomechanics of joint contracture. *Biorheology*, **17**, 95.

48. Akeson W.H., Woo S.L.-Y., Amiel D., *et al.* (1974). Biomechanical and biochemical changes in the periarticular connective tissue during contracture development in the immobilised rabbit knee. *Connect. Tissue Res.*, **2**, 315.

49. Akeson W.H., Amiel D., Mechanic G.L., *et al.* (1977). Collagen cross-linking alterations in joint contractures: change in the reducible cross-links in periarticular connective tissue collagen after nine weeks of immobilization. *Connect. Tissue Res.*, **5**, 15.

50. Lavigne A.B., Watkins R.P. (1973). Preliminary results on immobilisation-induced stiffness of monkey knee joints and posterior capsule. In *Perspectives in Biomedical Engineering* (Kenedi R.M. ed.). London: Macmillan.

51. Woo S.L.-Y., Mathews J.V., Akeson W.H., *et al.* (1975). Connective tissue response to immobility. Correlative study of biomechanical and biochemical measurements of normal and immobilised rabbit knees. *Arthritis Rheum.*, **18**, 257.
52. Peacock E.E. (1966). Some biochemical and biophysical aspects of joint stiffness: role of collagen synthesis as opposed to altered molecular bonding. *Ann. Surg.*, **164**.
53. Evans E.B., Eggers G.W.N., Butler J.K., *et al.* (1960). Experimental immobilisation and remobilisation of rat knee joints. *J. Bone Joint Surg.*, **42A**, 737.
54. Akeson W.H. (1961). An experimental study of joint stiffness. *J. Bone Joint Surg.*, **43A**, 1022.
55. Akeson W.H., Amiel D., Abel M.F., *et al.* (1987). Effects of immobilisation on joints. *Clin. Orthop.*, **219**, 28.
56. Akeson W.H., Woo S.L.-Y., Amiel D., *et al.* (1973). The connective tissue response to immobility: biochemical changes in periarticular connective tissue of the immobilised rabbit knee. *Clin. Orthop.*, **93**, 356.
57. Weeks P.M., Wray R.C. (1973). *Management of Acute Hand Injuries.* St. Louis: CV Mosby.
58. Strickland J.W. (1987). Biological basis for hand splinting. In *Hand Splinting: Principles and Methods* (Fess E.E., Phillips C.A. eds.). St. Louis: CV Mosby.
59. Noyes F.R. (1977). Functional properties of knee ligaments and alterations induced by immobilisation. A correlative and histological study in primates. *Clin. Orthop.*, **123**, 210.
60. Amiel D., Woo S.L.-Y., Harwood F.L. (1982). The effect of immobilisation on collagen turnover in connective tissue: a biochemical–biomechanical correlation. *Acta Orthop. Scand.*, **53**, 325.
61. Woo S.L.-Y. (1985). Functional adaptation and homeostasis of bone, tendons, and ligaments. In *Biomechanics: Current Interdisciplinary Research* (Perren S.M., Schneider E. eds.). Boston: Martinus Nijhoff, p. 73.
62. Woo S.L.-Y. (1986). Biomechanics of tendons and ligaments. In *Frontiers in Biomechanics* (Schmid-Schonbein G.W., Woo S.L.-Y., Zweifach B.W. eds.). New York: Springer-Verlag.
63. Frank C., Amiel D., Woo S.L.-Y., *et al.* (1985). Normal ligament properties and ligament healing. *Clin. Orthop.*, **196**, 15.
64. Enneking W.F., Horowitz M. (1972). The intra-articular effects of immobilisation on the human knee. *J. Bone Joint Surg.*, **54A**, 973.
65. Finsterbush A., Friedman B. (1975). Reversibility of joint changes produced by immobilisation in rabbits. *Clin. Orthop.*, **111**, 290.
66. Buckwalter J.A., Maynaud J.A., Vailas A.C. (1987). Skeletal fibrous tissues: tendon, joint capsule, and ligament. In *Scientific Basis of Orthopaedics.* 2nd edn. (Albright J.A., Brand R.A. eds.). London: Edward Arnold.
67. Catto M.E. (1985). Healing (repair) and hypertrophy. In *Muir's Textbook of Pathology.* 12th edn. (Anderson J.R. ed.). London: Edward Arnold.
68. Wilhelm D.L. (1977). Inflammation and healing. In *Pathology* Vol. 1 (Anderson W.A.D., Kissarie J.M. eds.). St. Louis: CV Mosby.
69. Hardy M.A. (1989). The biology of scar formation. *Phys. Ther.*, **69**, 1014.
70. Brand P.W. (1985). *Clinical Mechanics of the Hand.* St Louis: CV Mosby.
71. Laseter G.F. (1983). Management of the stiff hand: a practical approach. *Orthop. Clin. North Am.*, **14**, 749.
72. Strickland J.W. (1989). Biological rationale, clinical application, and results of early motion following flexor tendon repair. *J. Hand Ther.*, April–June, 71.
73. Gelberman R.H., Woo S.L.-Y. (1989). The physiological basis for application of controlled stress in the rehabilitation of flexor tendon injuries. *J. Hand Ther.*, April–June, 66.
74. Mathews P. (1979). The pathology of flexor tendon repair. *Hand*, **11**, 233.

75. Lundborg G.N., Rank F. (1987). Tendon healing: intrinsic mechanisms. In *Tendon Surgery in the Hand* (Hunter J.M., Schneider L.H., Mackin E.J. eds.). St Louis: CV Mosby.

76. Woo S.L.-Y., Buckwalter J.A. (1988). *Injury and Repair of the Musculoskeletal Soft Tissues*. Illinois: American Academy of Orthopaedic Surgeons.

77. Gelberman R.H., Manske P.R. (1987). Effect of early motion on the tendon healing process: experimental studies. In *Tendon Surgery in the Hand* (Hunter J.M., Schneider L.H., Mackin E.J. eds.). St Louis: CV Mosby.

73. Buckwalter J.A., Maynard J.A., Vailas A.C. (1987). Skeletal fibrous tissues: tendons, joint capsule and ligament. In *The Scientific Basis of Orthopaedics* 2nd edn. (Albright J.A., Brand R.A. eds.). California: Appleton and Lange.

79. Tardieu C., Huet de la Tour E., Bret M.D., *et al.* (1982). Muscle hypoextensibility in children with cerebral palsy: I. Clinical and experimental observations. *Dev. Med. Child Neurol.*, **63**, 97.

80. Corrigan B., Maitland G.D. (1983). *Practical Orthopaedic Medicine*. London: Butterworth.

81. Adams J.C. (1976). *Outline of Orthopaedics* 8th edn. Edinburgh: Churchill Livingstone.

82. Frankel V., Nordin M. (1980). *Basic Biomechanics of the Skeletal System*. Philadelphia: Lea and Febiger.

83. Hooley C.J., McCrum N.G., Cohen R.E. (1980). The visco-elastic deformation of tendon. *J. Biomech.*, **13**, 521.

84. McCarter R.J.M., Nabarro F.R.N., Wyndham C.H. (1971). Reversibility of the passive length–tension relation in mammalian skeletal muscle. *Arch. Int. Physiol. Biochem.*, **79**, 469.

85. Pradas M.M., Calleja R.D. (1990). Nonlinear viscoelastic behaviour of the flexor tendon of the human hand. *J. Biomech.*, **23**, 773.

86. Schwerdt H., Constantinesco A., Chambron J. (1980). Dynamic viscoelastic behaviour of the human tendon in vitro. *J. Biomech.*, **13**, 913.

87. Taylor D.C., Dalton J.D., Seaber A.V., *et al.* (1990). Viscoelastic properties of muscle-tendon units. The biomechanical effect of stretching. *Am. J. Sports Med.*, **18**, 300.

88. Gwinn T., Cronan C., Douglas C., *et al.* (1988). The acute affects of stretching on mechanical function of the ankle joint complex. Unpublished paper.

89. Kottke F.J., Parley D.L., Ptak R.A. (1966). The rationale for prolonged stretching for correction of shortening of connective tissue. *Arch. Phys. Med. Rehabil.*, **47**, 345.

90. Kottke F.J. (1971). Therapeutic exercise. In *Handbook of Physical Medicine and Rehabilitation* 2nd edn. (Krusen F.H., Kottke F.J., Ellwood P.M. eds.). Philadelphia: WB Saunders.

91. Sapega A.A., Quedenfield T.C., Moyer R.A., *et al.* (1981). Biophysical factors in range-of-motion exercises. *Physician Sports Med.*, **9**, 57.

92. Binkley J. (1989). Overview of ligament and tendon structure and mechanics: implications for clinical practice. *Physiother. Can.*, **41**, 24.

93. Hepburn G.R. (1987). Case studies: contracture and stiff joint management with dynasplint. *J. Orthop. Sports Phys. Ther.*, **8**, 498.

94. McKay-Lyons M. (1989). Low-load, prolonged stretch in treatment of elbow flexion contractures secondary to head trauma: a case report. *Phys. Ther.*, **69**, 292.

95. Bohannon R.W. (1984). Effect of repeated eight-minute muscle loading on the angle of straight-leg raising. *Phys. Ther.*, **64**, 491.

96. Riddle D. (1986). Case study: a treatment approach for a resistant knee extension contracture. *J. Orthop. Sports Phys. Ther.*, **7**, 159.

97. Carr J., Shepherd R.B. (1987). *A Motor Relearning Programme for Stroke*. 2nd edn. London: Heinemann.
98. Carr J., Shepherd R.B. (1987). A motor learning model for rehabilitation. In *Movement Science: Foundations for Physical Therapy in Rehabilitation* (Carr J.H., Shepherd R.B. eds.). London: Heinemann.
99. Michelsson J.E., Riska E.B. (1979). The effect of temporary exercising of a joint during an immobilisation period: an experimental study on rabbits. *Clin. Orthop.*, **144**, 321.
100. Woo S.L.-Y., Gelberman R.H., Cobb N.G., *et al.* (1981). The importance of controlled passive mobilisation on flexor tendon healing. *Acta Orthop. Scand.*, **52**, 615.
101. Gelberman R.H., Botte M.J., Spiegelman J.J., *et al.* (1986). The excursion and deformation of repaired flexor tendons treated with protected early motion. *J. Hand Surg.*, **11A**, 106.
102. Gelberman R.H., Vande Berg J.S., Lundborg G.N., *et al.* (1983). Flexor tendon healing and restoration of the gliding surface. An ultrastructural study in dogs. *J. Bone Joint Surg.*, **65A**, 70.
103. Strickland J.W. (1980). Digital function following flexor tendon repair in Zone II: a comparison of immobilisation and controlled passive motion techniques. *J. Hand Surg.*, **5**, 537.
104. Edinburg M., Widgerow A.D., Biddulph S.L. (1987). Early post-operative mobilisation of flexor tendon injuries using a modification of the Kleinert technique. *J. Hand Surg.*, **12A**, 34.
105. Browne E.Z., Ribik G.A. (1989). Early dynamic splinting for extensor tendon injuries. *J. Hand Surg.*, **14A**, 72.
106. Small J.O., Brennen M.D., Colville J. (1989). Early active mobilisation following flexor tendon repair in Zone II. *J. Hand Surg.*, **14B**, 383.
107. Chow J.A., Thomas L.J., Douelle S. (1987). A combined regimen of controlled motion following flexor tendon repair in 'no mans land'. *Plast. Reconstr. Surg.*, **79**, 447.
108. Chow J.A., Thomas L.J., Dovelle S. (1988). Controlled motion rehabilitation after flexor tendon repair and grafting. *J. Bone Joint Surg.*, **70B**, 591.
109. Stone R.G., Spencer E.L., Alonquist E.E. (1989). An evaluation of early motion management following primary flexor tendon repair: zones 1–3. *J. Hand Ther.*, Oct–Dec, 223.
110. Stegink Jansen G.W., Minerbo G. (1990). A comparison between early dynamically controlled mobilisation and immobilisation after flexor tendon repair in zone 2 of the hand: preliminary results. *J. Hand Ther.*, Jan–March, 20.
111. Hernandez A., Velasco F., Rivas A., Preciado A. (1967). Preliminary report on early mobilisation for the rehabilitation of flexor tendons. *Plast. Reconstr. Surg.*, **40**, 354.
112. Campbell D.T. and Stanley J.C. (1966). *Experimental and Quasi-experimental Designs for Research*. Chicago: Rand-McNally.
113. Cook T.D., Campbell D.T. (1979). *Quasi-experimentation. Design and Analysis Issues for Field Settings*. Boston: Houghton Mifflin.
114. Nielsen P.T., Rechnagel K., Nielsen S.-E. (1988). No effect of continuous passive motion after arthroplasty of the knee. *Acta Orthop. Scand.*, **59**, 580.
115. Coutts R.D., Toth C., Kaita J.H. (1984). The role of continuous passive motion in the rehabilitation of the total knee patient. In *Total Knee Arthroplasty* (Hungerford D.S., Krackow K.A., Kenna R.V. eds.). Baltimore: Williams and Wilkins, p. 126.
116. Romness D.W., Rand J.A. (1988). The role of continuous passive motion following total knee arthroplasty. *Clin. Orthop.*, **226**, 34.

117. Gose J.C. (1987). Continuous passive motion in the postoperative treatment of patients with total knee replacement. A retrospective study. *Phys. Ther.*, **67**, 39.

118. Laupattarakasem W. (1988). Short term continuous passive motion. *J. Bone Joint Surg.*, **70B**, 802.

119. Greene W.B. (1983). Use of continuous slow passive motion in the post-operative rehabilitation of difficult paediatric knee and elbow problems. *J. Pediatr. Orthop.*, **3**, 419.

120. Bunker T.D., Potter B., Barton N.J. (1989). Continuous passive motion following flexor tendon repair. *J. Hand Surg.*, **14B**, 406.

121. Stap L.J. (1986). Continuous passive motion in the treatment of knee flexion contractures. A case report. *Phys. Ther.*, **66**, 1720.

122. Basso D.M., Knapp L. (1987). Comparison to two continuous passive motion protocols for patients with total knee implants. *Phys. Ther.*, **67**, 360.

123. Evans R.E., Clark C.R., Miller B.A. (1986). Daily use of continuous passive motion in rehabilitation of total knee replacements: 10 vs 20 hours. *Phys. Ther.*, **66**, 803.

124. Harms M., Engstrom B. (1991). Continuous passive motion as an adjunct to treatment in the physiotherapy management of the total knee arthroplasty patient. *Physiotherapy*, **77**, 301.

125. Ritter M.A., Gandolf V.S., Holston K.S. (1989). Continuous passive motion versus physical therapy in total knee arthroplasty. *Clin. Orthop.*, **244**, 239.

126. Maloney W.J., Schurman D.J., Hangen D. (1990). The influence of continuous passive motion on the outcome of total knee arthroplasty. *Clin. Orthop.*, **256**, 162.

127. Johnson D.P. (1990). The effect of continuous passive motion on wound-healing and joint motion after knee arthroplasty. *J. Bone Joint Surg.*, **72A**, 421.

128. Davies S. (1991). Effects of continuous passive movement and plaster of paris after internal fixation of the ankle. *Physiotherapy*, **77**, 516.

129. Williams P.E. (1988). Effects of intermittent stretch on immobilised muscle. *Ann. Rheum. Dis.*, **47**, 1014.

130. Williams P.E. (1990). Use of intermittent stretch in the prevention of serial sarcomere loss in immobilised muscle. *Ann. Rheum. Dis.*, **49**, 316.

131. Williams P.E. (1989). Adaptation of muscles to changes in functional length. In *Motor Disturbances: Mechanisms and Implications for Therapy* (Torode M., Balnave R. eds.). Sydney: Cumberland College of Health Sciences, p. 47.

132. Ada L., Canning C. (1990). Anticipating and avoiding muscle shortening. In *Key Issues in Neurological Physiotherapy* (Ada L., Canning C. eds.). London Butterworth-Heinemann.

133. Scott J.A., Donovan W.H. (1981). The prevention of shoulder pain and contracture in the acute tetraplegia patient. *Paraplegia*, **19**, 313.

134. Sullivan T., Conine T.A., Goodman M., *et al.* (1988). Serial casting to prevent equinus in acute traumatic head injury. *Physiother. Can.*, **40**, 346.

135. Connie T.A., Sullivan T., Mackie T., *et al.* (1990). Effect of serial casting for the prevention of equinus in patients with acute head injury. *Arch. Phys. Med. Rehabil.*, **71**, 310.

136. Kent H., Hershler C., Conine T.A., *et al.* (1990). Case-control study of lower extremity serial casting in adult patients with head injury. *Physiother. Can.*, **42**, 189.

137. Booth B.J., Doyle M., Montgomery J. (1983). Serial casting for the management of spasticity in the head-injured adult. *Phys. Ther.*, **63**, 1960.

138. Light K.E., Nuzik S., Personius W., *et al.* (1984). Low-load prolonged stretch vs high-load brief stretch in treating knee contractures. *Phys. Ther.*, **64**, 330.

139. Bohannon R.W. (1985). Devices for increasing passive ankle dorsiflexion at home. Suggestion from the field. *Phys. Ther.*, **65**, 1521.
140. Bohannon R.W., Chavis D., Larkin P., *et al.* (1985). Effectiveness of repeated prolonged loading for increasing flexion in knees demonstrating post-operative stiffness. A clinical report. *Phys. Ther.*, **65**, 494.
141. Bohannon R.W., Larkin P.A. (1985). Passive ankle dorsiflexion increases in patients after a regimen of tilt table–wedge board standing. A clinical report. *Phys. Ther.*, **65**, 1676.
142. Dubuc W.E., Bohannon R.W. (1985). Device for stretching the hamstring muscles. Suggestion from the field. *Phys. Ther.*, **65**, 352.
143. Greene D.P., McCoy H. (1979). Turnbuckle orthotic correction of elbow-flexion contractures after acute injuries. *J. Bone Joint Surg.*, **61A**, 1092.
144. Fess E.E., Phillips C.A., eds. (1987). *Hand Splinting. Principles and Methods.* St Louis: CV Mosby.
145. MacKay-Lyons M. (1989). Low-load, prolonged stretch in the treatment of elbow flexion contractures secondary to head trauma: a case report. *Phys. Ther.*, **69**, 292.
146. Hepburn G.R. (1987). Case studies: contracture and stiff joint management with Dynasplint. *J. Orthop. Sp. Phys. Ther.*, April, 498.
147. Moseley A.M. (1991). The effect of serial casting on calf muscle length: a controlled trial. *Proceedings of the Eleventh International Congress of the World Confederation for Physical Therapy*, London, 998.
148. Maitland G.D. (1986). *Vertebral Manipulation*, 5th edn. London: Butterworth.
149. Olson V.L. (1987). Evaluation of joint mobilisation treatment. A method. *Phys. Ther.*, **67**, 351.

Chapter 6

Human Strength Adaptations – Implications for Therapy

ROB HERBERT

INTRODUCTION

A minimum level of strength is necessary for the performance of everyday motor tasks. This level is, however, not fixed; strength requirements of people will fluctuate over their lifetimes, and even from day to day. To a degree the neuromuscular system is capable of accommodating to these fluctuations, so that when it is regularly required to produce muscle tensions that approach or exceed its existing capacity, its tension generating abilities increase – the person becomes stronger. Conversely, when the system experiences a decrease demand for the production of large muscle tensions (when it experiences relative disuse), its tension generating abilities decrease, and the person may become weak.

The weakness that is brought about by neuromuscular adaptations to disuse can impede the performance of motor tasks, and it is primarily for this reason that physiotherapists try to prevent the development of disuse weakness, or to restore normal strength once disuse weakness has developed. Effective therapeutic intervention is only possible because of the adaptability of the tension generating abilities of the neuromuscular system. When weakness occurs in response to disuse it can usually be reversed simply by placing sufficient demands for tension production on the neuromuscular system. To this end, physiotherapists often augment the demands made on the neuromuscular system, usually with voluntary exercise or with electrical stimulation.

The aims of this chapter are to provide a broad overview of what is known about strength adaptations in humans, and to consider some of the implications of this information for clinical practice. However, because the literature in this area is extensive the scope of the chapter has been limited in a few ways. Firstly, an attempt has been made to discuss only those topics that have the most relevance to clinical practice. As a result, only passing mention will be made of research into strength-related adaptations in animals. Secondly, because of the focus on issues relevant to clinical practice, the parameter of primary interest will be the ability of people to actively generate muscle tension for the performance of everyday motor tasks. In this context, it will be most useful to define *strength* as the ability to voluntarily

142

generate active muscle tension (or torque), and *weakness* will be used to mean an inability to voluntarily generate the active muscle tension (or torque) that is necessary for the performance of everyday motor tasks. No specific consideration will be given here to the strength adaptations of people with disease or injury of the neuromuscular system, such as peripheral nerve lesions, stroke or muscular dystrophy. Instead, the discussion will centre on adaptations to increased or decreased use caused by circumstances such as cast immobilization, prolonged bed rest, and strength training, in people with 'intact' neuromuscular systems.

The chapter is divided into three sections. The first section will outline the nature of the problem of disuse weakness, and it will conclude with an evaluation of the effectiveness of some strategies that are widely used to prevent disuse weakness. The second section will consider how muscles respond to increased use, particularly the increased use brought about by structured strength training programmes. The main aim of this second section will be to define the parameters of a strength training programme that bring about the greatest strength increases in the shortest period of time. The last section will consider a particular case of strength training; namely, strength training for people with disuse weakness. Specifically, this section will address the questions of whether people with disuse weakness respond to training in the same way as people who are not weak, and whether strength training programmes facilitate the recovery of strength in people with disuse weakness.

DISUSE WEAKNESS

Relative disuse of muscles commonly occurs when joints are immobilized by casts or other fixation devices, during long periods of inactivity or confinement to bed, or in the presence of pain that is exacerbated by muscle contraction. In these situations, the neuromuscular system experiences relatively small demands for tension production, and it may respond by becoming less able to generate muscle tension. The magnitude of the strength losses associated with various periods of cast immobilization have been illustrated in Figure 6.1, using data compiled from the literature (see Table 6.1 and references 1–11 for details). This figure shows that cast immobilization is associated with a progressive decrease in strength. The reported strength losses associated with more than a couple of weeks of immobilization are quite substantial; in some cases they may be sufficient to impede the performance of everyday motor tasks. Moreover, it is apparent that when immobilization follows injury the resulting decreases in strength tend to be even greater than decreases in strength that occur in response to immobilization alone. No doubt this is due, at least in part, to the difficulty in generating strong muscle contractions in the presence of pain or joint effusion (see Chapter 2).

TABLE 6.1

Strength losses associated with cast immobilization

Reference number	n	Voluntary exercise?	Injury?	Comparison	Isometric/dynamic	Muscle group	Duration (days)	Mean loss of strength
1	8	–	+	Contralateral	Isometric	Thumb flexors	42	55%
2	10	–	–	Pre-disuse	Isometric	Knee extensors	9	13%
2	10	+	–	Pre-disuse	Isometric	Knee extensors	9	9%
3	5	–	–	Pre-disuse	Isometric	Thumb adductors	35	56%
3	6	–	–	Pre-disuse	Isometric	Thumb abductors	35	22%
4	1	–	–	Pre-disuse	Isometric	Grip	30	44%
5	10	–	–	Pre-disuse	Dynamic	Knee extensors	14	27%
5	10	–	–	Pre-disuse	Dynamic	Knee flexors	14	16%
5	10	–	–	Pre-disuse	Dynamic	Ankle pflexors	14	33%
5	10	+	–	Pre-disuse	Dynamic	Ankle dflexors	14	25%
5	10	+	–	Pre-disuse	Dynamic	Knee extensors	14	24%
5	10	+	–	Pre-disuse	Dynamic	Knee flexors	14	–4%
5	10	+	–	Pre-disuse	Dynamic	Ankle pflexors	14	19%
5	10	+	–	Pre-disuse	Dynamic	Ankle dflexors	14	8%
6	10	+	+	Pre-disuse	Isometric	Knee extensors	42	58%
7	7	–	–	Pre-disuse	Dynamic	Elbow extensors	35–42	41%
7	7	–	–	Pre-disuse	Isometric	Grip	7	0%
8	7	–	–	Pre-disuse	Isometric	Pinch	7	1%
8	7	–	–	Pre-disuse	Isometric	Thumb adductors	7	15%
8	30	–	+	Contralateral	Isometric	Grip	21–49	75%
8	30	–	+	Contralateral	Isometric	Pinch	21–49	50%
9	11	–	–	Pre-disuse	Isometric	Knee extensors	14	12%
9	11	–	–	Pre-disuse	Isometric	Knee extensors	14	28%
9	11	+	–	Pre-disuse	Dynamic	Knee extensors	14	29%
9	11	+	–	Pre-disuse	Isometric	Knee extensors	14	3%
9	11	+	–	Pre-disuse	Isometric	Knee extensors	14	–16%
9	11	+	–	Pre-disuse	Dynamic	Knee extensors	14	–14%
9	11	–	+	?	Isometric	Knee extensors	14–49	55%
9	11	–	+	?	Dynamic	Knee extensors	14–49	43%
9	11	+	+	?	Isometric	Knee extensors	14–49	45%
9	11	+	+	?	Dynamic	Knee extensors	14–49	74%
10	7	–	+	Pre-disuse	Isometric	Knee extensors	14–49	80%
11	8	?	+	Contralateral	Isometric	Knee extensors	27–43	58%

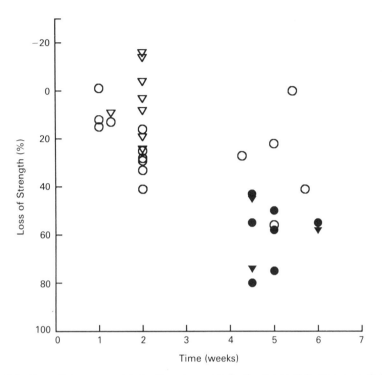

Figure 6.1 *Strength losses associated with various periods of cast immobilization, compiled from data reported in the literature (see Table 6.1 for details). Circles denote immobilization without concurrent voluntary exercise, triangles denote immobilization with concurrent voluntary exercise. Filled symbols indicate that immobilization followed injury.*

The strength losses associated with bed rest are broadly comparable in magnitude to those induced by cast immobilization[12–14] (Table 6.2). Thus, as with cast immobilization, long periods of bed rest may cause sufficient weakness to impede motor performance, particularly in frail people whose strength before confinement to bed may have only marginally exceeded that necessary for function.

Mechanisms

There are a number of morphological and physiological changes that could bring about the decreases in strength that accompany disuse. One such change is atrophy, or a decrease in the cross-sectional area of the muscle or muscle fibres.[6,7,12,15–20] (Atrophy is often also used to mean a decrease in muscle weight. In this chapter, however, atrophy will be used to mean only decreases in muscle cross-sectional area.) Atrophy would be expected to contribute to strength losses because the cross-sectional area of muscle is directly related to its intrinsic ability to generate tension[21,22] (i.e. to its ability

TABLE 6.2

Strength losses associated with prolonged bed rest

Reference number	n	Voluntary exercise?	Injury?	Comparison	Isometric/dynamic	Muscle group	Duration (days)	Mean loss of strength
12	9	—	—	Pre-disuse	Dynamic	Ankle pflexors	35	27%
12	9	—	—	Pre-disuse	Dynamic	Ankle pflexors	35	25%
12	9	—	—	Pre-disuse	Dynamic	Ankle dflexors	35	10%
13	15	—	—	Pre-disuse	Dynamic	Ankle pflexors	35	24%
13	15	—	—	Pre-disuse	Dynamic	Ankle pflexors	35	26%
13	15	—	—	Pre-disuse	Dynamic	Ankle dflexors	35	8%
13	15	—	—	Pre-disuse	Dynamic	Knee flexors	35	8%
13	15	—	—	Pre-disuse	Dynamic	Knee extensors	35	19%
13	15	—	—	Pre-disuse	Dynamic	Elbow flexors	35	7%
14	7	—	—	Pre-disuse	Dynamic	Knee extensors	30	19%
14	7	—	—	Pre-disuse	Dynamic	Knee flexors	30	6%
14	3	—	—	Pre-disuse	Dynamic	Knee extensors	30	19%
14	3	—	—	Pre-disuse	Dynamic	Knee flexors	30	9%

to generate tension when tetanically stimulated). The effect of atrophy is, therefore, to bring about a proportional decrease in the muscle's ability to generate tension.

Animal studies have described the time course of atrophy that accompanies disuse.[23–25] These studies show that, after a short threshold period during which no atrophy occurs, muscle cross-sectional area decreases, initially very rapidly and then progressively more slowly. To date very few studies have documented the time course of atrophy of immobilized human muscles. However, it is clear from studies on humans[6,7,15–17,19] that the extent of atrophy associated with a given period of cast immobilization is quite variable. For example, 5 weeks of cast immobilization has been variably reported as being accompanied by no decrease in muscle cross-sectional area,[16] or losses of cross-sectional area of 25 to 30%.[7,16]

The variability of the extent of atrophy reported in these studies suggests that the duration of immobilization is not the only factor that determines the degree of atrophy that occurs when muscles are immobilized. Another factor that is likely to be important is the length (or joint angle) at which the muscle is immobilized. Animal studies have demonstrated that when muscles are immobilized in a lengthened position they atrophy much less than when they are immobilized in a shortened position.[26] Although little research has systematically examined the effect of position of immobilization on atrophy in human muscles, some authors have noted that immobilization of the knee in extension is associated with a greater atrophy of the knee extensor muscles than of the knee flexor muscles.[16] This observation provides some support for the idea that some human muscles, like animal muscles, atrophy less when they are immobilized in a lengthened position.[27]

Other factors also determine the degree of atrophy that occurs with disuse. For example, some authors have suggested that when immobilization follows injury the atrophic response may be accelerated.[28,29] It is apparent also that different muscles,[25,30–38] and even different parts of the same muscle,[16,39] may respond to disuse with differing degrees of atrophy. This may be related to differences in muscle fibre type compositions, muscle architecture, or the demands for tension production that the muscles experienced prior to disuse.

In studies in which both the atrophy and the strength losses associated with disuse have been reported, the losses of strength have been consistently greater than the decreases in muscle cross-sectional area.[6,7,12,21] This is thought to indicate that, while atrophy is almost certainly a major cause of much disuse weakness, it is not the only mechanism by which losses of strength occur. Animal studies indicate that immobilization, and particularly immobilization in a shortened position, may be associated with a decrease in the muscle specific tension (the maximum tension the muscle can produce divided by muscle cross-sectional area; see reference 27 for review). This may be related to the observation that when muscles are immobilized in a shortened position, the normal orderly arrangement of myofibrils may become disrupted.[40,41] Whatever the explanation, the result is that the remaining muscle tissue appears to become less able to generate tension.

It is also possible that immobilization causes strength losses because it brings about a decrease in the ability to fully activate muscles. The findings of a small number of electromyographic studies[1,3,11,42] suggest that the losses of strength observed following short periods of immobilization may be partly due to a decreased ability to voluntarily activate muscles. In contrast, a study on subjects who demonstrated weakness at 1–5 years following injury and disuse found that all subjects were able to fully activate their muscles.[21] Perhaps the decreased ability to activate muscles following a period of disuse is a transient phenomenon.

To summarize this section, disuse causes progressive decreases in strength. The decrease in strength results at least partly from an adaptive decrease in muscle cross-sectional area, the extent of which appears to depend on the position in which the muscle was immobilized. Strength losses may also be attributable to a decrease in the remaining muscle's intrinsic ability to generate tension, and to a decreased ability to voluntarily activate the muscle.

PREVENTION OF DISUSE WEAKNESS

Physiotherapists are interested in preventing disuse weakness because it can impede the performance of motor tasks. In order to prevent disuse weakness they often prescribe voluntary (usually isometric) exercise to be performed during the period of disuse. The rationale is that, if disuse weakness occurs because muscles are deprived of the high tension contractions they usually experience in the performance of everyday movements, then the provision of high tension contractions in the form of voluntary exercise will retard the development of disuse weakness.

While there appears to be little doubt that the high tension muscle contractions performed during everyday movements provide a stimulus that regulates muscle strength, it is possible that the particular way in which muscles are required to contract during the performance of isometric exercises does not provide the mechanical stimulus necessary for the prevention of disuse weakness. That is, it is conceivable that typical exercise programmes consisting of small numbers of high tension isometric muscle contractions might not adequately provide the mechanical stimulus necessary to prevent disuse weakness. It may be important, for example, that the muscles experience lengthening or shortening while they are generating tension, as they do during the performance of everyday movement.[27] This would not (and could not) be a feature of the isometric exercise performed by a person whose muscles were immobilized in a plaster cast. Also, it is possible that the large number of low tension contractions that muscles normally experience during the performance of everyday movements represent a necessary part of the stimulus that maintains muscle strength. In fact, animal studies have shown that even very low tension muscle contractions can retard the development of disuse weakness in animals,[43–45] suggesting that low intensity muscle contractions have an important role in the regulation of muscle strength. Yet most exercise programmes designed to prevent post-immobilization weakness utilize only infrequent, high tension contractions, and these may be interspaced with long periods of little or no muscle activity. If either of these rather speculative hypotheses was true (i.e. if the necessary stimulus required lengthening and shortening contractions, or large numbers of low tension contractions) then isometric exercise might be expected to be ineffective in preventing disuse weakness. In fact, in studies in which muscle strength has been measured before and after a period of immobilization, substantial strength losses have almost invariably been observed, even when the subjects have performed isometric exercises throughout the period of immobilization[2,5,6,9,15] (see Figure 6.1 and Table 6.1). In the light of these observations it is reasonable to question whether isometric exercise during a period of cast immobilization is capable of retarding the development of disuse weakness or atrophy.

Evidence of a preventive effect of isometric exercise is elusive, because only a small number of relevant studies have used strong experimental designs[46,47] and because in some cases those studies that have used strong experimental designs have not reported their results in sufficient detail. Of those studies that have utilized random allocation of subjects to exercise and control groups, three studies have used non-injured people as subjects.[2,5,9] In all three studies the group that performed isometric exercise demonstrated consistently smaller strength losses than the group that did not perform exercise, and in some cases this difference between groups was very substantial. For example, in one study,[9] the mean decrease in strength of the group which performed isometric exercise was about 30% less than the mean decrease in strength of the group that did not receive isometric exercise. However, two of these studies failed to perform between-group statistical comparisons,[2,9] and in the third study[5] the necessary comparisons were apparently made but not reported. When statistical tests are performed on the data reported in these studies, they show that the differences between exercised and non-exercised groups are significant for all three measures of strength reported in one study[9], and for two of the four measures of strength reported in another study.[5]

In the remaining study, insufficient data were reported on which to perform statistical tests. Together these studies provide substantial support for the idea that isometric exercise can retard strength losses in cast immobilized muscles of non-injured subjects.

Only two studies that have used experimental designs have looked at the effect of isometric exercise on preventing the strength losses associated with immobilization after injury.[9,48] In one widely cited study,[48] soccer players who had sustained ligamentous injuries to the knee were immobilized in long-leg plaster casts for a number of weeks. They were allocated either to a group that performed no exercise during the period of immobilization, or to a group that performed isometric exercises. At the end of the period of immobilization the mean muscle strengths of the two groups were statistically indistinguishable. Consequently the authors concluded that there was no effect of isometric exercise. However, an examination of the data shows that the results of over two-thirds of the subjects in the control and exercise groups were not reported, presumably because these subjects were unable to complete the study. Also, it is possible (in the absence of any data on statistical power) that the failure of this study to show an effect of exercise could be due to the high probability of a statistical (type II) error.[49] In the light of these threats to internal and statistical validity, this study cannot be considered to provide strong evidence of a lack of effect of isometric exercise.

In the only other randomized study,[9] twenty-one people who were immobilized in a long leg cast as a result of lower limb injury or disease were randomly allocated to groups which performed either no exercise or ten, 10-second isometric contractions of their quadriceps every waking hour. After the period of immobilization it was found that the group that performed isometric exercise had a significantly greater mean strength, as measured by the 10RM, than the group that did not receive exercise (see p. 153 for a definition of the 10RM). The exercise group also exhibited greater isometric strengths, but the difference in isometric strength was not statistically significant. This study represents the strongest evidence to date of the ability of isometric exercise to retard post-immobilization weakness of injured subjects.

An alternative or additional method for preventing disuse weakness is the use of electrical stimulation. Electrical stimulation can be used to activate muscles either in the absence of, or when superimposed upon, concurrent voluntary efforts to activate the muscle. There are now quite a few studies that have investigated the effectiveness of electrical stimulation in preventing post-immobilization weakness,[5,6,10,48,50-53] and these studies permit some tentative conclusions to be drawn. There is evidence that the use of electrical stimulation alone is no more effective than the use of voluntary exercise alone in preventing losses of strength during the period of immobilization that follows knee reconstruction surgery.[53] On the other hand, when electrical stimulation is superimposed on voluntary muscle contractions, the resulting strength losses can be significantly less than those that occur with voluntary exercise alone.[6] A provisional interpretation of this literature might be that the use of electrical stimulation may be more effective (or perhaps it may only be effective) if it is accompanied by concurrent voluntary exercise. However, the conclusions from such a small number of studies cannot be considered definitive – further research is needed to replicate these findings and to determine their generality.

STRENGTH TRAINING

Often physiotherapists need to increase the strength of people with normal strength (i.e. of people who do not have disuse weakness). This is usually necessary when a person needs to use some muscle groups to generate unusually large tensions in order to compensate for weakness in other muscle groups. For example, spinal cord-injured people who are unable to generate tension with lower limb muscle groups may need to be able to generate large tensions with upper limb muscle groups in order to perform activities of daily living. Likewise, amputees may need to be able to generate large tensions with proximal muscle groups to compensate for the lost ability to actively produce muscle tension more distally. In these cases, muscles of normal strength need to become stronger. The extensive literature on the training response of normal subjects provides some guidelines for the prescription of exercise for this group of people.

It has become almost universally accepted that, in order to induce increases in strength, strength training programmes must utilize high tension muscle contractions. McDonagh and Davies[54] concluded in their review on strength training that a threshold load of about 66% of the maximum exists, below which dynamic strength training will not induce increases in isometric strength. The training response to isometric exercise also appears to exhibit a threshold tension, below which a strength training effect will not occur.[54] This threshold may be somewhat lower than that for dynamic exercise, because substantial strength increases have been noted even when subjects train isometrically at 25%[55] or 30%[56,57] of their maximum. However, in the studies that reported these findings subjects performed weekly maximal contractions so that training intensities could be adjusted. It may simply be that it was these weekly maximal contractions, and not the low intensity contractions used in training, that were responsible for the observed strength increases associated with low intensity training programmes. The existence of a training intensity threshold means that if the training load is not sufficiently high there will be little or no increase in strength with training. That is, if subjects are not required to produce sufficiently high tension muscle contractions they will not become stronger.

It is clear also that the training threshold is a function of strength.[58] That is, people who are able to produce large muscle tensions need to train at higher muscle tensions than people with less strength. Consequently, as a person's strength increases with training, the muscle tensions they are required to produce in their training programme (i.e. the loads they are required to lift) should increase at least in proportion with increases in strength in order to maintain a suprathreshold training stimulus. An effective training programme must therefore utilize progressive, high resistance loads.

The strength increases that occur in response to progressive high resistance exercise can be remarkably rapid. Increases in strength of up to 3% per day have been documented.[54,59] The magnitude of the observed strength increases are similar in males and females when the strength increases are expressed in terms of percentage of pretraining strength.[60] Also, it appears that elderly people respond to strength training in much the same way as younger people. The strength increases (relative to pretraining strength) associated with strength training in elderly people are broadly comparable in their magnitude with those of younger people.[59]

Mechanisms

Increases in muscle strength have been shown to be accompanied by increases in muscle cross-sectional area, or hypertrophy.[7,18,59–66] With training, there is a parallel addition of myofilaments to myofibrils, causing the muscle fibres, and hence whole muscles, to become larger in cross-section.[18] We have already seen that the intrinsic tension-generating ability of muscles is related to muscle cross-sectional area. It is clear, therefore, that hypertrophy is one mechanism by which muscle strength can be increased with training. However, the training-induced increases in muscle cross-sectional area are generally smaller than the accompanying increases in strength. For example, in one study[59] a 117% increase in dynamic strength was accompanied by only a 10% increase in muscle cross-sectional area; i.e. in this study the strength increases were much greater than those produced by hypertrophy alone.[7,60–65] Moreover, studies that compare pre- and post-training electrically-evoked muscle torques demonstrate that the intrinsic tension-generating ability of muscle increases much less than the observed increase in strength.[56,57,67,68] These observations have been taken to indicate that training-induced strength increases are not brought about solely by adaptations of the muscle itself.

The strength increases that occur over and above those due to changes in muscle have been attributed to a motor learning effect (sometimes called 'neural adaptations' in the strength training literature). It appears that strength training brings about strength increases partly because, with training, subjects learn to more completely or more effectively recruit motor units, or to more effectively utilize the tension muscles can produce.[54,63,69–72] The evidence for this is threefold. Firstly, some investigators have observed EMG changes with training that are suggestive of alterations in motor unit recruitment.[73,74] (On the other hand, there is some evidence that some subjects can accomplish huge strength gains with little muscle hypertrophy even if they are capable of fully activating their muscles before the onset of a training programme.[68,75] This suggests that, at least in some cases, neither hypertrophy nor an increased ability to activate muscles is responsible for the strength increases that accompany strength training.) A second line of evidence comes from observations of a training effect on the non-exercised contralateral limbs of subjects who undergo unilateral exercise.[71] The most likely explanation of the observed contralateral strength increases would appear to be that subjects learn to activate muscles more completely, or that they learn muscle activation strategies that enable them to lift heavy loads more effectively,[63] and that these learned activation patterns can also be utilized by the contralateral limb. Thirdly, many authors have documented a very specific training response (see Specificity of training, p. 156), and it is difficult to attribute such specific responses to muscular adaptations. Instead, specific training responses are easily explained by the learning of muscle activation strategies or muscle synergies that do not generalize well to other contexts. Taken together, these three lines of evidence provide support for the existence of neural mechanisms mediating strength increases that accompany training.

Optimal training intensities

An issue of importance to physiotherapists is the training parameters, and particularly the training intensity,[76] that brings about the greatest strength increases in the shortest period of time. Identification of the optimal training intensity can help physiotherapists to design optimally effective training programmes. The literature that has examined the optimal training intensity is extensive (see references 54, 76 and 77 for reviews) but few studies have systematically and independently manipulated training parameters, so synthesis of the literature is difficult. In this section, I will consider in turn the optimal training intensities for dynamic and isometric exercise.

In one widely used dynamic training protocol,[78,79] subjects are required to repeatedly lift and lower a weight until they are unable to continue lifting. This constitutes one 'set' of lifts, and the number of lifts in a set is called the 'repetition number'. The size of the load is measured by the number of times the subject can lift the weight without resting. Hence the maximum load a person can lift ten times without resting is their '10RM' (or ten 'repetition maximum') load. The effect of varying training load (and therefore the number of repetitions), and of varying the number of sets, has been investigated by Berger.[80] He trained subjects with loads of 2RM, 6RM and 10RM, and with one, two or three sets of lifts, three times a week for 12 weeks. At the end of the training programme he found that subjects achieved slightly greater increases in strength when they trained with the 6RM load than when they trained with the 2RM or 10RM loads (Figure 6.2). Three sets of lifts produced greater strength gains than those associated with either one or two sets of lifts. In a later study, Berger[81] narrowed the range of optimal training loads, this time training with one set only, to between 3RM and 9RM. Another study, which used a smaller sample size, has also examined the effect of varying the load in a similar training protocol. This study found no difference between the training response to three sets of 2–3RM or 5–6RM or 9–10RM loads.[82] A harmonious interpretation might be that (at least when subjects train with this protocol, repeatedly lifting a constant load until they can no longer lift it) there is little difference in the effectiveness of training within a range of high, submaximal loads, but that a reasonable estimate of the optimal training load is between the 3RM and 9RM loads. It would seem, therefore, that when physiotherapists train muscle strength using this dynamic training protocol, and where practical and safety considerations permit, the training load should be somewhere between 3 and 9RM.

It is common practice, when prescribing dynamic training programmes, to require that subjects continue lifting the training weight until they are unable to perform any further lifts. Thus the person training with a 6RM load will perform six lifts, and the person training with the 10RM load will perform ten lifts. In training programmes such as these, subjects are substantially fatigued during training. Recently it has been shown that fatiguing contractions are a necessary component of an optimally effective strength training programme.[83] Subjects were randomly allocated either to a group which performed a traditional training protocol utilizing substantially fatiguing contractions, or to a group which performed the same number of

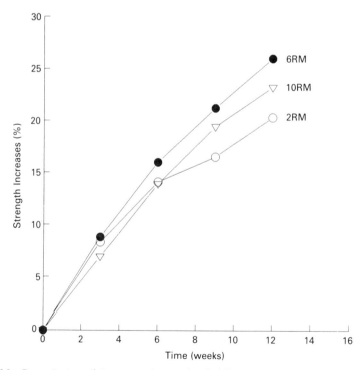

Figure 6.2   Dynamic strength increases associated with different training intensities. Graph shows increases in dynamic (1RM bench-press) strength produced by bench-press training with the 2RM, 6RM and 10RM loads. The greatest strength increases occurred with the 6RM training load. (Redrawn with permission from Berger R. (1962). Res. Q. Exerc. Sport., **33**, 168.)

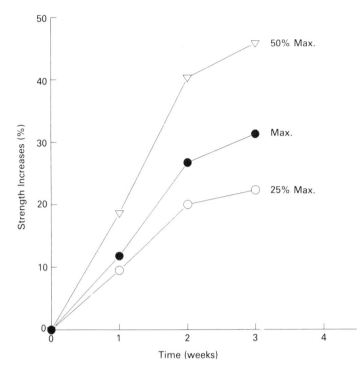

*Figure 6.3   Isometric strength increases associated with different training intensities. Graph shows increases in isometric strength produced by training at 25% and 50% of the maximum voluntary isometric torque, and with repeated maximal contractions. The greatest strength increases occurred when subjects trained at 50% of their maximum isometric torque. (Redrawn with permission from Szeto G., Strauss G.R., De Domenico G., et al. (1989). Aust. J. Physiother., **35**, 210.)*

lifts at the same relative load but with less fatigue (achieved by resting between lifts). Subjects who trained with fatiguing contractions achieved about a 50% greater mean increase in strength than subjects who trained with rests between contractions, indicating that the fatigue associated with traditional high-intensity training programmes augments the training stimulus.

The optimal isometric training intensity has received less attention than the optimal dynamic training intensity. One of the best studies to date compared the effect of training with three sets of ten, 5-second isometric contractions at either 25% or 50% of a maximal contraction, or with a maximal effort on each contraction.[55] Subjects trained on 5 days each week for 3 weeks. At the end of the training period, subjects who trained at 50% of maximum had substantially greater strength increases (mean increase of 46%) than those who trained with maximum contractions (mean increase of 31%) or those who trained at 25% of maximum (mean increase of 22%; Figure 6.3). This indicates that the optimal isometric training intensity, like the optimal dynamic training intensity, is submaximal.

An alternative to training muscle strength with voluntary dynamic or isometric exercise is to use electrically stimulated muscle contractions. A substantial body of literature has investigated whether the strength training stimulus can be better provided by electrical stimulation of muscles than by voluntary exercise (for reviews see references 71 and 84). If electrical stimulation is to be of value to physiotherapists training muscles of normal strength, electrically-induced muscle contractions must be more effective in increasing muscle strength than voluntary exercise alone. It has been argued on theoretical grounds that electrical stimulation could be associated with an enhanced training response because it involves different patterns of motor unit recruitment to voluntary exercise.[84] Unlike voluntary contractions, electrically stimulated muscle contractions involve a preferential recruitment of large, fast-twitch motor units, and it is conceivable that this recruitment pattern could generate a more marked training response than voluntary muscle contractions.[85] However, the evidence from studies on people without disuse weakness indicates that the electrical stimulation of muscle contractions, with or without concurrent voluntary muscle contractions, may be capable of producing strength increases that are comparable in magnitude, but no greater than, those produced by voluntary exercise alone.[71,83,86,87] On the basis of these studies it would seem that electrical stimulation, at least as it is currently administered, has little to offer in the training of people who are capable of voluntarily activating their muscles.

Specificity of training

The preceding discussion has considered the optimal parameters of training in terms of how much tension muscles should be required to produce in order to maximize the training response, and whether training is better provided by voluntary or electrically-induced contractions. Equally important is the mode of exercise that will produce the greatest increases in strength. Some authors have argued that the pattern of resistance provided by lifting free weights does not provide an optimal training stimulus because it is only at one point in range that muscles are required to contract at the required intensity. These authors argue that variable resistance and accommodating resistance (or isokinetic) machines provide a training stimulus that is superior to that provided by lifting free weights, because these machines require that the muscles generate the required intensity contraction throughout the full range of the movements. However, studies that have compared the strength training effects of these different modes of exercise have provided little convincing evidence of the superiority of any training machine.[88,89] At present there is no strong evidence to support the superiority of accommodating resistance or variable resistance machines over the more traditional free weights for strength training.

The decision of which mode of exercise is optimum therefore need not be particularly influenced by consideration of whether resistance should be provided by free weights, accommodating resistance, variable resistance, or any other means. The overriding issue should be how the muscles can be made to generate tension in a manner that most closely resembles the manner in which they are required to produce tension for task performance. This is because the training response is often quite specific to the mode of training employed. With training, the neuromuscular system will tend to become better at generating tension for actions that resemble the muscle actions employed in training, but not necessarily for muscle actions that are dissimilar to those used in training.

It is possible to identify at least four aspects of strength training specificity.[63,69] Firstly and most obviously, strength training effects are largely muscle specific. If one muscle group is required to regularly generate high tension contractions then predominantly those muscles, and not other muscles, will become stronger. (One exception, as we have already seen, is the contralateral training effect that occurs with high intensity strength training.) A second facet of the specificity of training effect is joint angle specificity. If a muscle is trained isometrically at one joint angle[90,91] or dynamically through a limited range of joint angles,[92,93] then the increases in strength that occur at that joint angle (or that range of joint angles) are likely to be less evident at other joint angles (Figure 6.4). Thirdly, the training response can be velocity specific. If muscles are trained at one velocity they may become stronger at that velocity, but less so at other velocities.[59,61,94–97] This means not only that slow concentric training may not substantially increase the ability to generate muscle tension at high velocities, but also that concentric training may fail to substantially change isometric (zero velocity) or eccentric (negative velocity) strength (Figure 6.5). In one study, for example, 12 weeks of concentric and eccentric training was shown to be accompanied by a mean increase of about 180% in dynamic strength, but the same subjects experienced only a mean 11% increase in isometric strength.[75] A fourth and particularly intriguing aspect of training specificity is postural specificity. An early study on training specificity[94] showed that, if subjects trained their elbow flexor strength by repeatedly generating large isometric elbow flexor torques in standing, then their ability to generate large isometric elbow flexor torques in standing increased, but there was relatively little increase in their ability to generate isometric elbow flexor torques in supine lying. This suggests that strength training involves not only learning to generate high muscle tensions, but also learning to generate high muscle tensions while simultaneously making appropriate task-specific postural adjustments and invoking appropriate motor synergies. To summarize, studies of training specificity have shown that the magnitude of the training response at the angles, velocities and postures in which training occurred are usually greater than, and never less than, the training response at other angles, velocities and postures.

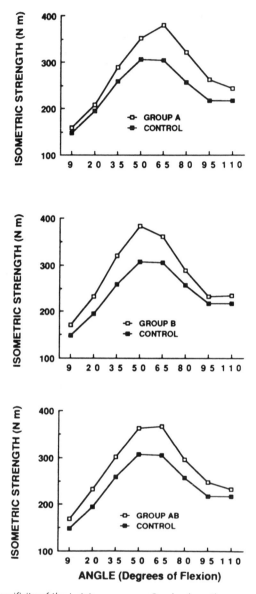

*Figure 6.4 Angle specificity of the training response. Graphs show the mean increase in isometric knee extensor strength at eight knee angles which accompanied three different training programmes. Subjects in group A (top panel) trained their knee extensors dynamically through the range of angles from 120° to 60°. Subjects in group B (middle panel) trained their knee extensors dynamically through the range of angles from 60° to 0°. Subjects in group AB (bottom panel) trained their knee extensors dynamically through the range of angles from 120° to 0°. The control group (shown in all panels) did not train. The greatest increases in strength occurred at the angles at which subjects trained. (Reproduced with permission from Graves J.E., Pollock M.L., Leggett S.H., et al. (1992). Med. Sci. Sports Exerc, **24**, 128.)*

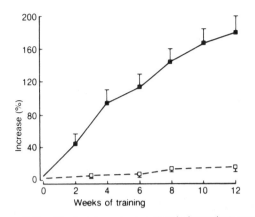

Figure 6.5   Velocity specificity of the training response. Graph shows the increases in mean isometric (open squares) and dynamic (6RM; filled squares) knee extensor strength of a group of subjects who trained their knee extensors dynamically for 12 weeks. Dynamic strength increased much more in response to dynamic training than did isometric strength. (Reproduced with permission from Rutherford O.M. (1988). *Sports Med.*, **5**, 196.)

The implication of this information for physiotherapists is as important as information on the optimal training intensity. It implies that when physiotherapists want to increase a person's strength so that the person can become better able to perform a motor task, training will be most effective if the muscles are trained in a manner that closely resembles the way in which they are required to generate tension for the performance of that task. Consider, for example, the paraplegic woman in Figure 6.6 who is unable to generate sufficient torque with the muscles of her upper limb and shoulder girdle to perform a floor to chair transfer. In this example, the physiotherapist could choose to institute a training programme that involved the woman flexing her shoulder in sitting against the load provided by weights and pulleys, or extending her elbow against the input arm of an isokinetic dynamometer. However, in these training programmes the muscles are being trained in other postures, at other joint angles, and probably at other velocities to those at which the muscles are required to produce tension during a floor to chair transfer. It may be then, that while the woman is getting stronger at lifting weights in sitting, or at pushing on the dynamometer arm, she may not be achieving the same magnitude increases in her ability to generate muscle torques for effecting a floor to chair transfer.

The implication is that strength training programmes aimed at improving task performance are likely to be most effective if they involve task specific muscle contraction types. One way of ensuring that the training programme is task specific is to practise the task itself, or a close approximation of the task.[63,98] For example, in the case of the paraplegic woman who is unable to perform the floor to chair transfer, training could initially consist of performing lifts off blocks of a low height (Figure 6.6b). In this case, if the training intensity is to be optimized, the block heights should be a little less

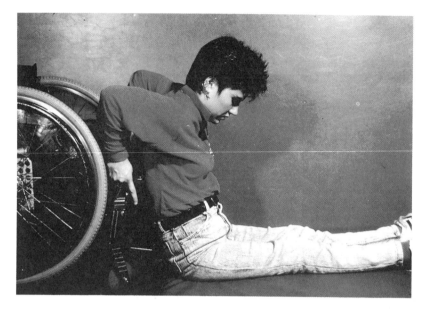

(a)

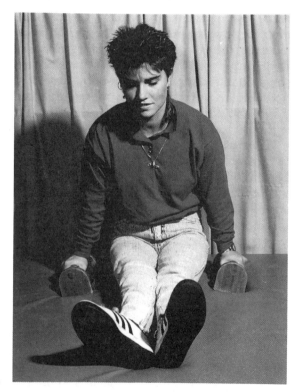

(b)

(c)

Figure 6.6   Task-specific training of strength. (a) This paraplegic woman is unable to perform a floor to wheelchair transfer. (b) She trains in a task-specific manner by practising lifting herself from blocks. (c) The height of the blocks is progressed so that she can only just successfully complete about six consecutive lifts. In this way the necessary training stimulus can be provided in a task-specific manner. (Photographs courtesy of Lisa Harvey, Royal Prince Henry Hospital, Sydney and the University of New South Wales Medical Illustrations Unit.)

than the maximal height from which the patient can effect a lift – say, a block height which they could successively negotiate just six times, a sort of functional 6RM load.[99] As the person becomes stronger, the block height could be progressively increased to maintain an optimal training intensity, by progressively increasing the height of the blocks (Figure 6.6c). In this way the optimal training parameters could be provided in a task-specific training programme, using the weight of the body to provide the training load. The muscles of the upper limb would be required to generate tension at approximately the joint angle and joint angular velocities, and in a similar body posture, as those angles, angular velocities and postures in which they are required to generate tension during a floor to chair transfer. Also, the person is learning to generate high tension muscle contractions at the same time as she is being required to make the postural adjustments that are required for the successful execution of a transfer.

For these reasons it would appear that the training (or practise) of *tasks*, rather than the traditional method of training muscle strength outside of the context of task performance, is likely to be a particularly effective way of increasing strength. But there are other advantages that could also stem from the practise of tasks for strength training. The most significant of these advantages is that the use of task practise obviates the need for physio-therapists to infer muscle actions from task kinematics. Contemporary approaches to strength training require that physiotherapists observe task performance and make decisions on the basis of these observations about which muscle groups are to be trained. Those muscles are sometimes then trained in isolation; i.e. joints other than those which the muscles cross are stabilized, and one joint only is required to move against an external load. However, at least in some circumstances it can be difficult to make reasonable inferences about muscle forces from observations of task kinematics.[100,101] It could be that the clinical process of deciding which muscles need to be strengthened on the basis of observations of task performance is prone to error, and that sometimes physiotherapists train the wrong muscle groups because they have not appreciated the true complexity of muscle actions. If physiotherapists train task performance, rather than training muscles in isolation, there is less of a requirement to infer muscle actions from task kinematics, and (as long as the subject is prevented from utilizing compensatory strategies) they can be relatively confident that the appropriate muscles are being trained.[98]

Strength training for people with disuse weakness

After a period of disuse, many people will have weakness that limits motor performance. Once the period of disuse has ended, and they resume or attempt to resume normal motor activity, strength usually recovers rapidly. Animal studies show that unrestrained cage activity following even quite prolonged disuse is accompanied by a rapid return of normal or near normal muscle tension-generating ability.[36,102,103] These studies demonstrate that, in animals, the return of normal motor activity provides a sufficient stimulus for the complete return of strength. In contrast, several papers have reported that even following long periods of time and intensive training programmes, many humans still demonstrate an incomplete recovery of strength following disuse.[21,104–106] Perhaps this apparent discrepancy between the findings of animal and human studies is related to the fact that humans are usually immobilized following injury, whereas the animals in these studies are not.

When disuse weakness appears to be limiting motor performance, physiotherapists often prescribe exercise to restore normal strength and to improve the person's ability to perform motor tasks. The implicit assumption underlying the rational prescription of exercise in this case is that a structured exercise programme can increase or hasten the recovery of strength, and that the recovery of strength will result in improved motor performance. Evidence for the reasonableness of this assumption comes from some clinical reports, and the widely made clinical observation, that when people with long-standing disuse weakness (say, disuse weakness of duration greater than 1 year) undergo structured strength training programmes they can experience large increases in strength.[21,106] That is, it seems clear that in this group of people strength training can facilitate the return of normal strength.

The situation is less clear when exercise is provided *soon after* the period of disuse, because then large strength increases are expected to occur as a result of the stimulus provided simply by the return to everyday motor activities. That is, even where structured strength training programmes are not provided, rapid increases in strength probably still occur in the period immediately following disuse. Consequently, it is not immediately obvious whether strength training induces greater or faster increases in strength than would otherwise be expected to occur. In order to ascertain confidently whether strength training is capable of facilitating recovery of strength soon after a period of disuse, it would be necessary to randomly allocate subjects to groups that receive either strength training or no strength training soon after the end of the period of disuse. Such a study has not yet been performed, perhaps because of ethical constraints, and so it is not possible to draw strong conclusions about the effectiveness, or lack of effectiveness, of strength training soon after disuse. A small number of studies have, however, compared the effectiveness of different strength training programmes, and these studies give some insights into the effect of strength training following disuse.

In one study,[107] 107 soccer players who had sustained knee injuries were immobilized in leg casts for between 3 and 7 weeks. After the period of immobilization each subject received progressive high resistance training with either dynamic, isometric or isokinetic exercise, or two bouts of about 6 minutes of one-legged cycling. Subjects trained three times each week for 4 weeks. At the end of the training period there was no significant difference in the strength of the different training groups (as measured by maximal isometric torque, maximal isokinetic torque, or 10RM). These findings are surprising, and they could mean one of several things. They could indicate that the training intensity threshold and the specificity of the training response are phenomena that apply only to the training of normal muscle, and not to muscles with disuse weakness. In this study even low intensity exercise such as one-legged cycling was accompanied by increases in strength, and the strength gains associated with all modes of training were statistically indistinguishable. Alternatively, this data could indicate that the everyday activity experienced by all subjects provided a powerful training stimulus, so that progressive high resistance training was insufficient to provide a training stimulus which was superior to that of everyday activity alone. A third possibility is that, in this study, the subject's strength could have been limited by pain or anxiety about re-injury, and the increases in strength exhibited by all groups could simply reflect a decrease in anxiety or pain. Lastly, it could be that the failure of this study to find a difference between training groups resulted from methodological problems, such as a bias in the allocation of subjects to groups or a lack of statistical power. Clearly, further research is needed to investigate these possibilities.

Other studies have also failed to find differences between the strength increases associated with various training programmes used to restore strength after immobilization. For example, no difference has been found between the training response to one or two sessions of training, three times each week after knee injury;[108] between one and two training sessions daily after menisectomy;[109,110] or between one and two training sessions daily after hip fracture.[111] Also, there appears to be no difference between the strength gains of subjects who receive early strength training (starting within 24 hours of cast removal) or late training (starting 2–3 weeks later), after immobilization for fractures of the radius.[8] One study found that dynamic training was slightly more effective than isometric training at increasing dynamic training after knee injury, but only after the sample size had been increased from 72 subjects[112] to 167 subjects.[113]

This failure of a number of studies to find differences in the training response to different training programmes begs the question of whether strength training soon after immobilization is capable of inducing strength increases over and above those increases that would occur in response to the performance of everyday activities alone. That is, it could simply be that the muscle activity associated with the performance of everyday motor tasks is sufficient to provide a powerful stimulus for increases of strength to occur, and that strength training programmes are unable to induce still greater increases in strength.

To my knowledge, the most compelling evidence of an additional training stimulus provided by voluntary exercise instituted soon after a period of disuse comes from a study on strength training after knee reconstruction.[114] In this study it was found that people who received electromyographic feedback about contraction intensity during training experienced significantly greater strength increases than subjects who exercised without biofeedback. This suggests that training with feedback aimed at maximizing contraction intensity can increase the training response. Perhaps more significantly, if it can be assumed that the strength increases of subjects who trained without biofeedback was not *less* than those strength increases that would occur with no training, then this study provides evidence that training with biofeedback is more effective than no training at increasing strength soon after immobilization.

A small number of well controlled studies have investigated the effectiveness of electrical stimulation when it is used for training muscle strength after disuse. In these studies, people whose knees had been immobilized after injury were randomly allocated to groups that received either voluntary exercise alone or electrical stimulation. The electrical stimulation was either superimposed upon voluntary muscle contractions,[115,116] or was provided in addition to voluntary muscle contractions,[117] or was unaccompanied by any voluntary contractions.[118] The findings of these studies are somewhat mixed: in one study,[118] electrical stimulation that was unaccompanied by voluntary muscle contractions was found to be more effective than voluntary muscle contractions alone, but another study[117] found no difference between electrical stimulation provided in addition to voluntary contractions, and voluntary muscle contractions alone. Of the studies that have compared the effectiveness of electrical stimulation superimposed on voluntary muscle contractions with voluntary muscle contractions alone, one found no difference,[116] and another found that electrical stimulation enhanced the training response.[118] Other authors have remarked upon the apparent inconsistencies of studies investigating the effect of electrical stimulation[83,84] and have suggested that the inconsistencies may reflect the different stimulation parameters used in different studies. Whatever the explanation, it is clear that further research is needed to investigate the effectiveness of training, both with voluntary exercise and with electrical stimulation, following periods of disuse.

SUMMARY

The neuromuscular system is capable of adapting its ability to generate large muscle tensions to the demands made upon it. When there is a decreased need for muscles to generate high tensions, as occurs with cast immobilization or prolonged bed rest, muscles may become weak. This is due at least in part to muscle atrophy, but it is probably also due to a decreased ability to activate muscles effectively. Physiotherapists often try to retard the development of immobilization-induced weakness with the use of voluntary isometric exercise or with electrical stimulation instigated during the period of immobilization. There is some empirical evidence that suggests that these strategies can prevent the development of weakness that accompanies immobilization following injury.

Studies on normal subjects (i.e. non-immobilized and uninjured subjects) indicate that progressive high resistance exercise can increase strength by inducing muscle hypertrophy and by increasing the ability to activate muscles in a way that generates large muscle tensions. The optimal intensity at which training can take place is slightly submaximal; for dynamic exercise the optimal training load appears to be between about 3RM and 9RM. Quite small numbers of contractions of this intensity are sufficient to induce large increases in muscle strength, particularly if training is structured so that subjects are fatigued at the end of each set of lifts. Importantly, the strength training response is specific to the way in which the muscle is trained; muscles that are trained at one velocity, angle or posture will tend to get stronger at that velocity, angle or posture, but the increase in strength at other velocities, angles and postures will tend to be smaller. This suggests that, if physiotherapists want to increase a person's strength to improve performance of a motor task, the most effective training may sometimes be carefully structured practise of the task itself.

Following a period of disuse, physiotherapists often institute strength training programmes to facilitate the restoration of normal muscle strength and to improve a person's ability to perform motor tasks. It seems that strength training does increase strength in people who demonstrate long-standing strength deficits following periods of disuse. However, there is some evidence that muscles with acute disuse weakness do not respond to training in the same way as muscles of normal strength. Perhaps this is because of the confounding effects of everyday motor activities, which may themselves provide a powerful stimulus for the restoration of strength following disuse. It is not yet certain whether strength training provided soon after a period of disuse, either in the form of voluntary muscle contractions or electrically stimulated muscle contractions, can induce a faster or more complete recovery of muscle strength than that provided by the performance of everyday activities alone.

ACKNOWLEDGEMENTS

I would like to thank Tom Gwinn and Louise Ada for their helpful comments on an earlier draft of this chapter.

REFERENCES

1. Duchateau J., Hainaut K. (1987). Electrical and mechanical changes in immobilised human muscle. *J. Appl. Physiol.*, **62**, 2168.
2. Rozier C.K., Elder J.D., Brown M. (1979). Prevention of atrophy by isometric exercise of a casted leg. *J. Sports Med.*, **19**, 191.
3. Sale D.G., McComas A.J., MacDougall J.D., *et al.* (1982). Neuromuscular adaptation in human thenar muscles following strength training and immobilisation. *J. Appl. Physiol.*, **53**, 419.
4. Hills W.L. (1973). Effects of immobilisation in the human forearm. *Arch. Phys. Med. Rehabil.*, **54**, 87.
5. Gould N., Donnermeyer D., Pope M., *et al.* (1982). Transcutaneous muscle stimulation as a method to retard disuse atrophy. *Clin. Orthop.*, **164**, 215.
6. Wigerstad-Lossing I., Grimby G., Jonsson T., *et al.* (1988). Effects of electrical muscle stimulation combined with voluntary muscle contractions after knee ligament surgery. *Med. Sci. Sports Exerc.*, **20**, 93.
7. MacDougall J.D., Elder G.C.B., Sale D.G., *et al.* (1980). Effects of strength training and immobilisation on human muscle fibres. *Eur. J. Appl. Physiol.*, **43**, 25.
8. Sykes K. (1988). *An investigation of the time course of recovery of muscle function after immobilisation.* Unpublished MSc thesis, University of Southampton.
9. Stillwell D.M., McLarren G.L., Gersten J.W. (1967). Atrophy of quadriceps muscle due to immobilisation of the lower extremity. *Arch. Phys. Med. Rehabil.*, **48**, 289.
10. Morrisey M.C., Brewster C.E., Shields C.L., *et al.* (1985). The effects of electrical stimulation on the quadriceps during postoperative knee immobilisation. *Am. J. Sports Med.*, **13**, 40.
11. Fuglsang-Frederiksen A., Scheel V. (1978). Transient decrease in number of motor units after immobilisation in man. *J. Neurol. Neurosurg. Psychiatry*, **41**, 924.
12. Le Blanc A., Gogia P., Schneider V. (1988). Calf muscle area and strength changes after five weeks of horizontal bed rest. *Am. J. Sports Med.*, **16**, 624.
13. Gogia P.P., Schneider V.S., Le Blanc A.D. (1988). Bed rest effect on extremity muscle torque in healthy men. *Arch. Phys. Med. Rehabil.*, **69**, 1030.

14. Dudley G.A., Duvoisin M.R., Convertino V.A., *et al*. (1989). Alterations of the *in vivo* torque–velocity relationship of human skeletal muscle following 30 days exposure to simulated microgravity. *Aviat. Space Environ. Med.*, **60**, 659.

15. Haggmark T., Jansson E., Eriksson E. (1981). Fiber type area and metabolic potential of the thigh muscle in man after knee surgery and immobilisation. *Int. J. Sports Med.*, **2**, 12.

16. Ingemann-Hansen T., Halkjaer-Kristensen J. (1980). Computerised tomographic determination of human thigh components. The effects of immobilisation in plaster and subsequent physical training. *Scand. J. Rehabil. Med.*, **12**, 27.

17. Gibson J.N.A., Halliday D., Morrison W.L. (1987). Decrease in human quadriceps muscle protein turnover consequent upon leg immobilisation. *Clin. Sci.*, **72**, 503.

18. McDougall J.D. (1984). Morphological changes in human skeletal muscle following strength training and immobilisation. In *Human Muscle Power* (Jones N.L., McCartney N., McComas A. J. eds.). Illinois: Human Kinetics.

19. Sargeant A.J., Davies C.T.M. (1977). The effect of disuse muscular atrophy on the forces generated in dynamic exercise. *Clin. Sci. Mol. Med.*, **53**, 183.

20. Hikida R.S., Gollnick P.D., Dudley G.A., *et al*. (1989). Structural and metabolic characteristics of human skeletal muscle following 30 days of simulated microgravity. *Aviat. Space Environ. Med.*, **60**, 664.

21. Rutherford O.M., Jones D.A., Round J.M. (1990). Long-lasting unilateral muscle wasting and weakness following injury and immobilisation. *Scand. J. Rehabil. Med.*, **22**, 33.

22. Close R.I. (1972). Dynamic properties of mammalian skeletal muscles. *Physiol. Rev.*, **52**, 129.

23. Tomanek R.J., Lund D.D. (1974). Degeneration of different types of skeletal muscle fibres. II. Immobilisation. *J. Anat.*, **118**, 531.

24. Appell H.J. (1986). Morphology of immobilised skeletal muscle and the effects of a pre- and post-immobilisation training programme. *Int. J. Sports Med.*, **7**, 6.

25. Thomasen D.B., Booth F.W. (1990). Atrophy of the soleus muscle by hind-limb unweighting. *J. Appl. Physiol.*, **68**, 1.

26. Spector S.A., Simard C.P., Fournier M., *et al*. (1982). Architectural alterations of rat hindlimb skeletal muscles immobilised at different lengths. *Exp. Neurol.*, **76**, 94.

27. St.-Pierre D., Gardiner P.F. (1987). The effect of immobilisation and exercise on muscle function: a review. *Physiother. Can.*, **39**, 24.

28. Halkjaer-Kristensen J., Ingemann-Hansen T. (1985). Wasting of human quadriceps muscle after knee ligament injuries. I. Anthropometrical consequences. *Scand. J. Rehabil. Med.*, **Suppl. 13**, 5.

29. Stokes M., Young A. (1984). The contribution of reflex inhibition to arthrogeneus muscle weakness. *Clin. Sci.*, **67**, 7.

30. Booth F.W. (1977). Time course of muscular atrophy during immobilisation of hindlimbs in rats. *J. Appl. Physiol.*, **43**, 656.

31. Loughna P., Goldspink G., Goldspink D.F. (1986). Effect of inactivity and passive stretch on protein turnover in phasic and postural rat muscles. *J. Appl. Physiol.*, **61**, 173.

32. Goldspink D.F., Goldspink G. (1986). The role of passive stretch in retarding muscle atrophy. In *Electrical Stimulation and Neuromuscular Disorders* (Nix W. A., Vrbova G. eds.). Berlin: Springer-Verlag.

33. Jaspers S.R. (1988). Effect of immobilisation on rat hindlimb muscles under non-weight-bearing conditions. *Muscle Nerve*, **11**, 458.

34. Shaw S.R., Zernicke R.F., Vailas A.C. (1987). Mechanical, morphological and biochemical adaptations of bone and muscle to hindlimb suspension and exercise. *J. Biomech.*, **20**, 225.
35. Fitts R.H., Metzger J.M., Riley D.A., *et al.* (1986). Models of disuse: a comparison of hindlimb suspension and immobilisation. *J. Appl. Physiol.*, **60**, 1946.
36. Fitts R.H., Brimmer C.J. (1985). Recovery in skeletal muscle contractile function after prolonged hindlimb immobilisation. *J. Appl. Physiol.*, **59**, 916.
37. Edgerton V.R., Barnard R.J., Peter J.B., *et al.* (1975). Properties of immobilised muscles of the Galago senegalensis. *Exp. Neurol.*, **46**, 115.
38. Witzmann F.A., Kim D.H., Fitts R.H. (1982). Hindlimb immobilisation: length–tension and contractile properties of skeletal muscle. *J. Appl. Physiol.*, **53**, 335.
39. Leiber R.L., Fridèn J.O., Hargens A.R., *et al.* (1988). Differential response of the dog quadriceps muscle to external skeletal fixation of the knee. *Muscle Nerve*, **11**, 193.
40. Cooper R.R. (1972). Alterations during immobilisation and regeneration of skeletal muscle in cats. *J. Bone Joint Surg.*, **54A**, 919.
41. Baker J.H., Matsumoto D.E. (1988). Adaptation of skeletal muscle to immobilisation in a shortened position. *Muscle Nerve*, **11**, 231.
42. Duchateau J., Hainaut K. (1990). Effects of immobilisation on contractile properties, recruitment and firing rates of human motor units. *J. Physiol.*, **422**, 55.
43. Thomasen D.B., Herrick R.E., Baldwin K.M. (1987). Activity influences on soleus muscle myosin during rodent hindlimb suspension. *J. Appl. Physiol.*, **63**, 138.
44. Michel R.N., Gardiner P.F. (1990). To what extent is hindlimb suspension a model of disuse? *Muscle Nerve*, **13**, 646.
45. Pierotti D.J., Roy R.R., Flores V., *et al.* (1990). Influence of 7 days of hindlimb suspension and intermittent weight support on rat muscle mechanical properties. *Aviat. Space Environ. Med.*, **61**, 205.
46. Campbell D.T., Stanley J.C. (1966). *Experimental and Quasi-experimental Designs for Research*. Chicago: Rand-McNally.
47. Cook T.D., Campbell D.T. (1979). *Quasi-experimentation. Design and Analysis for Field Settings*. Boston: Houghton Mifflin.
48. Halkjaer-Kristensen J., Ingemann-Hansen T. (1985). Wasting of the human quadriceps muscle after knee ligament injuries. IV. Dynamic and static muscle function. *Scand. J. Rehabil. Med.*, **Suppl. 13**, 29.
49. Cohen J. (1988). *Statistical Power Analysis*, 2nd edn. New York: Academic Press.
50. Duvoisin M.R., Convertino V.A., Buchanan P., *et al.* (1989). Characteristics and preliminary observations of the influence of electromyostimulation on the size and function of human skeletal muscle during 30 days of simulated microgravity. *Aviat. Space Environ. Med.*, **60**, 671.
51. Gould N., Donnermeyer D., Gammon G.G. (1983). Transcutaneous muscle stimulation to retard disuse atrophy after open meniscectomy. *Clin. Orthop.*, **178**, 190.
52. Eriksson E., Haggmark T. (1979). Comparison of isometric muscle training and electrical stimulation supplementing isometric muscle training in the recovery after major knee ligament surgery. A preliminary report. *Am. J. Sports Med.*, **7**, 169.
53. Sisk T.D., Stralka S.W., Deering M.B., *et al.* (1987). Effect of electrical stimulation on quadriceps strength after reconstructive surgery of the anterior cruciate ligaments. *Am. J. Sports Med.*, **15**, 215.
54. McDonagh M.J.N., Davies C.T.M. (1984). Adaptive response to mammalian skeletal muscle to exercise with high loads. *Eur. J. Appl. Physiol.*, **52**, 139.

55. Szeto G., Strauss G.R., De Domenico G., *et al.* (1989). The effect of training intensity on voluntary isometric strength improvement. *Aust. J. Physiother.* **35**, 210.
56. Young K., McDonagh M.J.N., Davies C.T.M. (1985). The effects of two forms of isometric training on the mechanical properties of the triceps surae in man. *Pflugers Arch.*, **405**, 384.
57. Davies C.T.M., Young K. (1984). Effect of training at 30 and 100% maximal isometric force (MVC) on the contractile properties of the triceps surae in man. *J. Physiol.*, **336**, 22P.
58. Enoka R.M. (1988). *Neuromechanical Basis of Kinesiology.* Illinois: Human Kinetics.
59. Frontera W.R., Meredith C.N., O'Reilly K.P. (1988). Strength conditioning in older men: skeletal muscle hypertrophy and improved function. *J. Appl. Physiol.*, **64**, 1038.
60. Cureton K.J., Collins M.A., Hill D.W., *et al.* (1988). Muscle hypertrophy in men and women. *Med. Sci. Sports Exerc.*, **20**, 338.
61. Jones D.A., Rutherford O.M. (1987). Human strength training: the effects of three different regimes and the nature of the resultant changes. *J. Physiol.*, **391**, 1.
62. Ikai M., Fukunaga T. (1970). A study on training effect on strength per unit cross-sectional area of muscle by means of ultrasonic measurement. *Int. Z. Agnew. Physiol.*, **28**, 173.
63. Rutherford O.M. (1988). Muscular co-ordination and strength training. Implications for injury rehabilitation. *Sports Med.*, **5**, 196.
64. Petersen S., Wessell J., Bagnall K., *et al.* (1990). Influence of concentric resistance training on concentric and eccentric strength. *Arch. Phys. Med. Rehabil.*, **71**, 101.
65. Young A., Stokes M., Round J.M., *et al.* (1983). The effect of high resistance training on the strength and cross-sectional area of the human quadriceps. *Eur. J. Clin. Invest.*, **13**, 411.
66. Lüthi J.M., Howald H., Claasen H., *et al.* (1986). Structural changes in skeletal muscle with heavy resistance exercise. *Int. J. Sports Med.*, **7**, 123.
67. McDonagh M.J.N., Hayward C.M., Davies C.T.M. (1983). Isometric training in human elbow flexor muscles. The effects on voluntary and electrically evoked forces. *J. Bone Joint Surg.*, **65B**, 355.
68. Ishida K. (1990). Changes in voluntary and electrically induced contractions during strength training and detraining. *Eur. J. Appl. Physiol.*, **60**, 244.
69. Sale D.G. (1987). Influence of exercise and training on motor unit activation. *Exerc. Sports Sci. Rev.*, **15**, 95.
70. Sale D.G. (1988). Neural adaptation to resistance training. *Med. Sci. Sports Exerc.*, **20**, S135.
71. Enoka R.M. (1988). Muscle strength and its development. New perspectives. *Sports Med.*, **6**, 146.
72. Komi P.V. (1986). How important is neural drive for strength and power development in human skeletal muscle? In *Biochemistry of Exercise VI* (Saltin B. ed.). Illinois: Human Kinetics.
73. Milner-Brown H.S., Stein R.B., Lee R.G. (1975). Synchronization of human motor units: possible roles of exercise and supraspinal reflexes. *Electroencephalogr. Clin. Neurophysiol.*, **38**, 245.
74. Cannon R.J., Cafarelli E. (1987). Neuromuscular adaptations to training. *J. Appl. Physiol.*, **63**, 2396.
75. Rutherford O.M., Jones D.A. (1986). The role of learning and co-ordination in strength training. *Eur. J. Appl. Physiol.*, **55**, 100.
76. Atha J. (1981). Strengthening muscle. *Exerc. Sports Sci. Rev.*, **9**, 1.
77. Clarke D.H. (1973). Adaptations in strength and muscular endurance resulting from exercise. *Exerc. Sports Sci. Rev.*, **1**, 73.

78. De Lorme T.L. (1945). Restoration of muscle power by heavy-resistance exercises. *J. Bone Joint Surg.*, **27**, 645.
79. De Lorme T.L., Watkins A.L. (1951). *Progressive resistance exercise: technic and medical application*. New York: Appleton-Century-Crofts.
80. Berger R. (1962). Effect of varied weight training programs on strength. *Res. Q. Exerc. Sport*, **33**, 168.
81. Berger R.A. (1962). Optimum repetitions for the development of strength. *Res. Q. Exerc. Sport*, **33**, 334.
82. O'Shea P. (1966). Effects of selected weight training programs on the development of strength and muscle hypertrophy. *Res. Q. Exerc. Sport*, **37**, 95.
83. Rooney K., Herbert R., Balnave R. Fatiguing contractions are a necessary component of an optimally effective strength training programme. Paper submitted for review.
84. Lloyd T., De Domenico G., Strauss G. R., *et al.* (1986). A review of the use of electro-motor stimulation in human muscles. *Aust. J. Physiother.*, **32**, 18.
85. Delitto A., Snyder-Mackler L. (1990). Two theories of muscle strength augmentation using percutaneous electrical stimulation. *Phys. Ther.*, **70**, 158.
86. Kramer J.F., Semple J.E. (1983). Comparison of selected strengthening techniques for normal quadriceps. *Physiother. Can.*, **35**, 300.
87. McMiken D.F., Todd-Smith M., Thompson C. (1983). Strengthening of human quadriceps muscles by cutaneous electrical stimulation. *Scand. J. Rehabil. Med.*, **15**, 25.
88. Kosmahl E.M., Mackarey P.J., Buntz S.E. (1989). Nautilus training system versus traditional weight training system. *J. Orthop. Sports Phys. Ther.*, **11**, 253.
89. Manning R.J., Graves J.E., Carpenter D.M. (1990). Constant vs variable resistance knee extension training. *Med. Sci. Sports Exerc.*, **22**, 397.
90. Thèpaut-Mathieu C., Van Hoeke J., Maton, B. (1988). Myoelectric and mechanical changes linked to length specificity during isometric training. *J. Appl. Physiol.*, **64**, 1500.
91. Kitai T.A., Sale D.G. (1989). Specificity of joint angle in isometric training. *Eur. J. Appl. Physiol.*, **58**, 744.
92. Graves J.E., Pollock M.L., Jones A.E., *et al.* (1989). Specificity of limited range of motion variable resistance training. *Med. Sci. Sports Exerc.*, **21**, 84.
93. Graves J.E., Pollock M.L., Leggett S.H., *et al.* (1992). Limited range of motion lumbar extension strength training. *Med. Sci. Sports Exerc.*, **24**, 128.
94. Rasch P.J., Morehouse L.E. (1957). Effect of static and dynamic exercises on muscular strength and hypertrophy. *J. Appl. Physiol.*, **11**, 29.
95. Kanehisa H., Miyashita M. (1983). Specificity of velocity in strength training. *Eur. J. Appl. Physiol.*, **52**, 104.
96. Duchateau J., Hainaut K. (1984). Isometric or dynamic training: differential effects on mechanical properties of a human muscle. *J. Appl. Physiol.*, **56**, 296.
97. Dons B., Bollerup K., Bonde-Petersen F., *et al.* (1979). The effect of weight-lifting exercise related to muscle fiber composition and muscle cross-sectional area in humans. *Eur. J. Appl. Physiol.*, **40**, 95.
98. Carr J., Shepherd R.B. (1987). A motor learning model for rehabilitation. In *Movement Science: Foundations for Physical Therapy in Rehabilitation* (Carr J., Shepherd R.B. eds.). London: Heinemann.
99. Gwinn T. (1988, 1989). Personal communication.
100. Winter D.A. (1985). Concerning the scientific bases for the diagnosis of pathological gait and rehabilitation protocols. *Physiother. Can.*, **37**, 245.
101. Zajac F.E., Gordon M.E. (1989). Determining muscle's force and action in multi-articular movement. *Exerc. Sports Sci. Rev.*, **17**, 187.
102. Booth F.W., Seider M.J. (1979). Recovery of skeletal muscle after 3 mo of hindlimb immobilisation in rats. *J. Appl. Physiol.*, **47**, 974.

103. Witzmann F.A., Kim D.H., Fitts R.H. (1982). Recovery time course in contractile function of fast and slow skeletal muscle after hindlimb immobilisation. *J. Appl. Physiol.*, **53**, 677.
104. Arvidsson I., Eriksson E., Haggmark T., *et al*. (1981). Isokinetic thigh muscle strength after ligament reconstruction in the knee joint: results from a 5–10 year follow-up after reconstructions of the anterior cruciate ligament in the knee joint. *Int. J. Sports Med.*, **2**, 7.
105. Singer K.P. (1986). The influence of unilateral electrical stimulation on motor unit activity patterns in atrophic human quadriceps. *Aust. J. Physiother.*, **32**, 31.
106. Grimby G., Gustafsson E., Peterson L., *et al*. (1980). Quadriceps function and training after knee ligament surgery. *Med. Sci. Sports Exerc.*, **12**, 70.
107. Ingemann-Hansen T., Halkjaer-Kristensen J. (1985). Physical training of the hypotrophic quadriceps muscle in man. I. The effects of different training programmes on muscular performance. *Scand. J. Rehabil. Med.*, **13**, 38.
108. Ingemann-Hansen T., Halkjaer-Kristensen J. (1983). Progressive resistance exercise training of the hypotrophic quadriceps muscle in man. The effects of morphology, size and function as well as the influence of duration of effort. *Scand. J. Rehabil. Med.*, **15**, 29.
109. Karumo I. (1977). Intensive physical therapy after menisectomy. *Ann. Chir. Gynaecol.*, 66: 41.
110. Karumo I., Rehunen S., Naveri H., Alho A. (1977). Red and white muscle fibres in menisectomy patients. Effects of postoperative physiotherapy. *Ann. Chir. Gynaecol.*, **66**, 164.
111. Karumo I. (1977). Recovery and rehabilitation of elderly subjects with femoral neck fractures. *Ann. Chir. Gynaecol.*, **66**, 170.
112. Zohn D.A., Leach R.E., Stryker W.S. (1964). A comparison of isometric and isotonic exercises of the quadriceps after injuries to the knee. *Arch. Phys. Med. Rehabil.*, **45**, 571.
113. Leach R.E., Stryker W.S., Zohn D.A. (1965). A comparative study of isometric and isotonic quadriceps exercise programs. *J. Bone Joint Surg.*, **47A**, 1421.
114. Draper V. (1990). Electromyographic biofeedback and recovery of quadriceps femoris muscle function following anterior cruciate ligament reconstruction. *Phys. Ther.*, **70**, 1.
115. Delitto A., Rose S.J., McKowen J.M. (1988). Electrical stimulation versus voluntary exercise in strengthening thigh musculature after anterior cruciate ligament surgery. *Phys. Ther.*, **68**, 660.
116. Lainey C.G., Walmsley R.P., Andrew G.M. (1983). Effectiveness of exercise alone versus exercise plus electrical stimulation in strengthening the quadriceps muscle. *Physiother. Can.*, **35**, 5.
117. Williams R.A., Morrissey M.C., Brewster C.E. (1986). The effect of electrical stimulation on quadriceps strength and thigh circumference in menisectomy. *J. Orthop. Sports Phys. Ther.*, **8**, 143.
118. Godfrey C.M., Jayawardena H., Quance T.A., *et al*. (1979). Comparison of electro-stimulation and isometric exercise in strengthening the quadriceps muscle. *Physiother. Can.*, **31**, 265.

Chapter 7

Promoting Effective Segmental Alignment

JENNY McCONNELL

INTRODUCTION

Patients with musculoskeletal problems mainly present to physiotherapists for the treatment of pain. Other symptoms, including instability and stiffness may be associated with pain, but are rarely reported by the patient, especially if there has been a gradual onset. The patient may have been able to compensate using other movements; for example, cervical spine stiffness, which should cause a problem when reversing a car, can often go undetected because the individual will turn the trunk to perform the task. It is usually only when the movement becomes painful that treatment is sought.

Faulty alignment and posture have been cited as not only perpetuating musculoskeletal problems but also as being the source of these problems.[1-3] Sahrmann claims that 'musculoskeletal pain syndromes are seldom caused by isolated precipitating events but are the consequences of habitual imbalances in the movement system'.[1] Other authors have indicated that poor neuromotor function contributes to various postural anomalies.[4,5] As a relationship exists between alignment, posture and musculoskeletal dysfunction, the key to managing and preventing musculoskeletal problems lies in the promotion of effective segmental alignment.

Posture was defined in 1947 by the American Academy of Orthopaedic Surgeons as: 'Relative alignment of the parts of the body. Good posture is that state of muscular and skeletal balance which protects the supporting structures of the body against injury or progressive deformity irrespective of the attitude (e.g. erect, lying, squatting or stooping) in which these structures are working or resting. Under such conditions the muscles will function most efficiently and the optimum positions are afforded for the thoracic and abdominal organs. Poor posture is a faulty relationship of the various parts of the body which produces increased strain on the supporting structures and in which there is less efficient balance of the body over its base of support'. This definition has since been adopted by numerous others.[5-7] However, objective criteria for normal posture are not in abundance and some criteria, particularly in dynamic situations such as lifting, are quite contradictory.[8-13]

The shortage of readily pertinent information for physiotherapists about key factors affording optimal segmental alignment, has resulted in difficulties in determining the appropriate management strategies for multifactorial chronic problems such as patellofemoral pain, supraspinatus tendonitis and low back pain. This chapter will, therefore, address the issue of promoting effective segmental alignment, by examining the following:

1 The importance of assessment to determine the symptom-producing structure or structures.
2 Methods of measuring outcome.
3 Treatment strategies for promoting effective segmental alignment. This will include the rationale for obtaining optimal joint loading, balanced muscle activity and adequate mobility of the neuromeningeal structures.

ASSESSMENT

Identification of the structure or structures causing the symptoms can sometimes be difficult, so a methodical approach should be adopted during the examination of a patient. It is not acceptable for physiotherapists to treat symptoms without having first determined the underlying dysfunctional mechanisms, because successful relief of pain during treatment can often mask more serious pathology. Identification of the underlying mechanisms involves questioning the patient about the location, nature, behaviour and onset of the symptoms. The patient's response to certain questions, especially in acute injuries, can often alert the clinician to the possible structure or structures involved. The physical examination, and in some situations further investigative procedures, will confirm the diagnosis. The following case example demonstrates the importance of careful history taking.

> *Case study 1:* A 51-year-old handyman presented with a swollen, painful right knee. Two weeks previously he had been kneeling while weeding. When he stood up and twisted to the right, he heard and felt a click in his knee, which was followed by immediate pain. He wriggled his leg and it clicked again. The knee clicked on two further occasions that day but it did not 'click back' after the last episode. The pain was most intense on the medial aspect of the knee and he was unable to straighten the knee fully. He was unable to sleep well at night, particularly if lying on his left side. His knee was aggravated by walking and there was no position in which the knee was comfortable.

The information gleaned from the history, which enables the physiotherapist to suspect a degenerate medial meniscal problem, is outlined below.

Firstly, a swollen knee that cannot be fully straightened indicates an internal derangement of the joint. Secondly, the sustained flexion position of kneeling is likely to damage the posterior horn of the meniscus. The structure is poorly vascularized and susceptible to injury, especially in the older age group when the fibrocartilage becomes more brittle. Thirdly, the presence of a click then a locking indicates a loose body or meniscal problem.

Fourthly, the presence of localized medial pain, which prevents the patient from sleeping on his opposite side at night, indicates a medial joint line problem.[14,15]

If a degenerative medial meniscal problem is suspected by the physiotherapist, the patient requires an immediate referral to a medical practitioner for an operative solution to his problem. The information gathered during the history taking is quite conclusive in determining the nature of the problem. Physical examination of the knee is unlikely to add substantially to the data gathered or to help confirm the diagnosis.

On some occasions, however, it may be difficult to ascertain exactly which structures have contributed to the injury because the mechanism of injury was vague. Case study 2 highlights the importance of the physical examination to differentiate the injured structure.

> Case study 2: A 20-year-old male, who was running during a touch football match, heard a loud crack in his left lower leg. He collapsed in immediate pain and thought he had been hit by a stray golf ball because golfers were practising at the other end of the oval. He was carried into the clinic complaining of medial ankle pain. On examination, he could not bear weight on the affected leg without severe pain and a feeling of instability. On resisted inversion and plantarflexion, severe pain was produced and the tibialis posterior tendon subluxed over the medial malleolus.

A provisional diagnosis of a rupture of the sheath of the tibialis posterior tendon was made and the patient referred to a medical practitioner. The provisional diagnosis was confirmed during a subsequent surgical repair.

The most likely injury resulting in the above symptoms would be rupture of the Achilles tendon. This would be unusual in a 20-year-old, the injury being more common in older subjects. However, rupture of the sheath of the tibialis posterior tendon is fairly rare, so it was only with specific muscle testing of each of the muscles with tendons running posterior to the medial malleolus, that the damaged structure was identified. This case study emphasized the importance of a thorough examination of the patient by the physiotherapist, based on anatomical knowledge, to determine the damaged structure.

Identification of the involved structure or structures, for the patient with chronic pain on the other hand, can often be a more difficult task, as the symptom manifestation site may be remote from the structures predisposing to the problem. Shoulder pain, as is depicted in Case study 3, is a good example.

> Case study 3: A 70-year-old female who had been treated for a frozen shoulder with shoulder mobilization and electrotherapy for 18 months presented with a still painful and stiff right shoulder, which she thought had plateaued in recovery. At night she was woken by the pain when lying on that shoulder. She was unable to fasten her bra from the back, brush her hair using that arm or carry a basket of washing. A previous cortisone injection into the shoulder had alleviated some of the pain but did not restore her movement.
> On examination the anterior and posterior aspects of the shoulder were tender on palpation, flexion was restricted to 120°, abduction to 100° and she

was only able to reach as far as the right posterior superior iliac spine with her hand behind her back. The upper fibres of the trapezius initiated arm raising, particularly through abduction. A marked thoracic kyphosis was observed. On passive movement, range of flexion was 20° greater than during active movement. Glenohumeral passive accessory movements were slightly, but not significantly, restricted compared with the other side. Thoracic spine accessory movements from T1–T7 were all markedly restricted.

Patients with frozen shoulder problems rarely complain of pain or even stiffness in their thoracic region, but gradual diminution of shoulder range is often caused by a lack of mobility in this area. It has been demonstrated in normal individuals that inner range unilateral shoulder flexion is accompanied by contralateral side flexion of the spine, while bilateral shoulder flexion induces spinal extension.[16,17] Therefore, a decrease in thoracic mobility or an increase in thoracic kyphosis will cause a decrease in shoulder range of movement.[18–20] An increased thoracic kyphosis may limit shoulder flexion by either decreasing the thoracic spine's contribution to motion, or by decreasing the scapulothoracic excursion. An increased thoracic kyphosis is said to abduct and rotate the scapula caudally which can result in proximal shoulder girdle muscle imbalance.[3] A downward rotated and protracted scapula could result in premature abutment of the humerus on the acromion and a reduced range of total shoulder flexion.[1] A downward and protracted scapula will also increase the resting length of some of the scapular stabilizers such as the lower trapezius muscle, which acts with the serratus anterior and upper trapezius to rotate the scapula in an upward direction to increase shoulder girdle stability beyond 90°.[21,22] A decrease in activity of an elongated lower trapezius will alter the balance in the scapular force couple and often result in an increase in the onset and amount of upper trapezius activity. This situation results in a completely altered arm elevation pattern and may further predispose the individual to an impingement problem, with compression of soft tissues between the acromion and the humeral head.

In the older age group where frozen shoulder conditions are more common, there is an increased incidence of both increased thoracic kyphosis and decreased thoracic mobility.[23] This significantly affects shoulder function. A study by Crawford and Jull[23] found that asymptomatic older subjects utilized almost all their available thoracic extension when they elevated their arms whereas younger subjects used only 50% of their available range. The available range of thoracic extension had decreased by 35% in the older age group, and kyphosis had increased by 36%. Shoulder flexion had, however, only decreased by 9%. Crawford and Jull suggested that subjects with pathological shoulder problems might be expected to demonstrate even greater decreases in thoracic mobility which would severely limit shoulder range.

With the above information in mind, the aims of treatment for the patient in Case study 3 would be to improve the thoracic spine mobility, particularly into extension and lateral flexion, to decrease the early initiation of upper trapezius activity during arm raising and to retrain lower trapezius and rotator cuff function.

To demonstrate the effect of treatment especially in the long term, the outcome needs to be measured. Rothstein has stated that 'without a scientific basis for the assessment (and measurement) process, physical therapists face the future as independent practitioners unable to communicate with one another, unable to document treatment efficacy, and unable to claim scientific credibility for the profession'.[24] The measurements used must be valid and reliable if the conclusions reached about the intervention are to be meaningful.[24,25]

MEASUREMENT OF OUTCOME

As most patients are concerned about their pain, a baseline measurement of the severity of pain at rest and during activity can be recorded on a Visual Analog Scale.[26] This can be used to monitor the effectiveness of intervention and any carry-over effect. Physiotherapists may also use this information to determine if there is a pattern emerging for particular conditions, which may help them refine intervention strategies.

To document change in function, objective measurements are required. Among the methods commonly used by physiotherapists are goniometers, spondylometers and tape measures to assess range of motion. Muscle performance can be quantified in a number of ways, including the measurement of muscle torque output using isokinetic dynamometers or the measurement of timing of muscle contraction, using EMG analysis. The inherent validity and reliability of any of these techniques will determine the usefulness of the measurement. Simple, valid and reliable approaches are more useful than complex costly methods of doubtful validity.

Some devices can be used for both measurement and treatment. Richardson and colleagues[27] have developed a pressure transducer to monitor changes in pressure between the trunk and a support surface (plinth, chair or wall) during stabilization exercises. Use of this biofeedback device has enhanced precision and control of patients during stabilization exercises for the lumbar and cervical spines.[27] The pressure biofeedback is designed to be readily used in the clinic and is useful for both assessment and training (Figure 7.1).

Dynamic segmental alignment can be recorded objectively to help the physiotherapist determine any dysfunctional elements during movement and to provide an objective measure of change. Numerous techniques of varying capability and cost as well as suitability are available.[28–31] Video-analysis of particular activities such as running, descending stairs or serving a tennis ball, using markers on anatomical landmarks, can provide useful information at relatively low cost. The data can be analysed at different levels of complexity depending upon the needs of the clinician; angular and

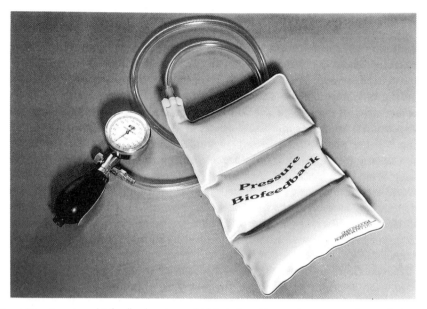

Figure 7.1   A pressure biofeedback system, which is designed to monitor pressure changes between the spine and an interface surface, will give an indication of the endurance capacity of the trunk and neck stabilizing muscles. The endurance capacity of the oblique muscles are monitored by the patient, as he attempts to keep the pressure a constant, 10 mmHg higher than the resting position for as long as possible. This becomes his baseline holding capacity. Improvement can, therefore, be monitored over time.

linear displacement can be measured directly, while velocity and acceleration data can be calculated.

A study by Ada and Westwood, using kinematic analysis of standing up at two different functional stages in subjects following stroke, demonstrated that improvement in the ability to stand up was related more to velocity than to angular displacement. The significance of this finding is that the key parameter, velocity, is not easily appreciated by an observer. Only with film or video analysis can these values be computed.[32]

TREATMENT STRATEGIES

Analysis of movements and the response of symptoms to movement will determine the treatment approach required for patients with a musculoskeletal disorder. The broad principle of treatment for most musculoskeletal problems are relatively straightforward. It is the *execution* of the treatment that can sometimes be difficult.

Treatment should be directed towards improving segmental alignment by: obtaining an optimal loading and mobility of the joint; gaining balanced muscle activity with appropriate synergist/stabilizer activity; and ensuring adequate mobility of the neuromeningeal structures.

Optimal loading of a joint

Promotion of optimal joint loading is a major aim of treatment in chronic musculoskeletal conditions because articular cartilage is nourished and maintained by evenly distributed, intermittent compression.[33–35] Optimal loading of a joint can be accomplished, to some extent, by increasing the surface area of contact of the joint, because, although contact pressure is proportional to the transmitted force, it is inversely proportional to the area.[36] If the physiotherapist can optimize the surface area of contact of a joint during treatment, then the force through the joint is distributed over a wider area. The force per unit area (i.e. pressure) will decrease and this should lead to a corresponding decrease in the pain. The process by which optimization of joint contact area can be promoted is given below using the patellofemoral joint as an example. The relevant anatomy and biomechanics of the joint are reviewed to help explain the rationale behind the method of optimization.

The functions of the patella are to link the divergent quadriceps muscle to a common tendon, to increase the extensor moment of the quadriceps muscle, to protect the tendon from compressive stress and to minimize stress concentration by transmitting forces evenly to the underlying bone.[37–42] Like all lower limb joints, the patellofemoral joint dissipates compressive stress by maximizing its surface area of contact.[43,44] With increasing knee flexion, when the compressive force increases (from $0.5 \times$ bodyweight for level walking to $7–8 \times$ body weight for squatting),[38,43,44] a greater proportion of the patella surface is in contact with the femur.[37,38,45–48]

The position of the patella relative to the trochlea of the femur is, to a large extent, controlled by the surrounding soft tissues.[38] Patients with patello-femoral symptoms, however, demonstrate a failure of the intricate balance of the soft tissue structures. The imbalance, due to various biomechanical faults, alters the distribution of the load to the undersurface of the patella which, ultimately, produces pain.[49–59] A major objective in treatment is to realign the patella with the trochlea of the femur so that the articulating surfaces are parallel to one another and the patella is midway between the two condyles when the knee is flexed to $20°$,[60–62] thus maximizing contact surface area. Maintaining the patella in this position should reduce pain immediately, enabling previously painful activities to be performed without pain.[60–62]

As the alignment of the patellofemoral joint is essentially determined by the surrounding soft tissues, the position of the patella can be changed by stretching the tight lateral retinacular structures and by changing the activation pattern of the vastus medialis obliquus (VMO) muscle. Stretching adaptively shortened retinacular tissues is achieved by a sustained low load to facilitate a permanent elongation of the tissues, utilizing the creep phenomenon, which occurs in viscoelastic material when a constant low load is applied.[63,64] Creep manifests itself over a period of time that may vary from

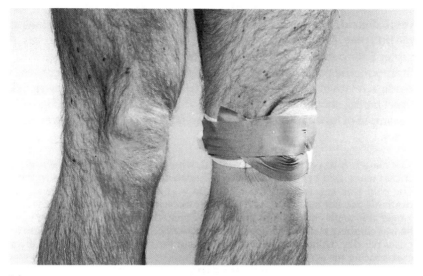

(a)

(b)

Figure 7.2   (a) Tape can be used to correct lateral tilt, lateral glide, anteroposterior tilt and external rotation of the patella. More than one component may need to be corrected to adequately align the patella with the trochlea. The tape should immediately reduce the patient's symptoms. (b) Once in a pain-free position, the patient is ready to commence training of the VMO. An EMG biofeedback is monitoring the patient's muscle activity.

several seconds to several days.[65] Consequently, tissue adaptations may come into play, so that permanent plastic deformation of tissues occurs. (See Chapter 5 for a more detailed explanation of the process.)

Strapping the patella with adhesive tape into an improved alignment will provide a relatively constant low load to the shortened retinacular tissue. Correct taping should always decrease symptoms immediately (Figure 7.2). An example of how taping the patella can facilitate recovery is given in Case study 4.

> *Case study 4:* A 16-year-old school student, who was a fast bowler in the A-grade cricket team, had been experiencing infrapatellar pain for the past 6 months. The pain was aggravated by bowling and was localized to the region underneath his patellar tendon, which would on occasions become puffy. He was no longer able to play cricket and the pain had recently started to bother him when he was climbing stairs at school. He was prescribed a regimen of 250 straight leg raises per day. Diligently, the boy carried out the treatment only to find it was exacerbating his symptoms.
>
> On examination, the pain was reproduced while he was weight bearing on an extended symptomatic knee. Extension overpressure of the tibiofemoral joint reproduced the symptoms passively.

The compromised structure was the fat pad, which was being irritated by the inferior pole of the patella being pulled posteriorly during end range extension manoeuvres of the knee.[66] The fat pad was unloaded with tape (Figure 7.3) and the patient was given a training programme that included specific training for the VMO. The patient's symptoms were dramatically reduced and he was able, after the initial treatment, to return to cricket. However, the tape failed when he was bowling, so the fat pad was no longer unloaded and the symptoms returned. Video analysis of his bowling action revealed that at delivery his right knee locked into extension, causing a posterior displacement of the inferior pole of the patella, which impinged the fat pad with each delivery. With modification to his bowling action, by getting him to land with a slightly flexed knee, his symptoms abated completely. As a bonus, the new bowling technique meant that he could bowl even faster.

Taping the patella has been shown to change significantly the lateral patellofemoral angle and lateral patellar displacement as measured radiologically.[67] In a study by Roberts,[67] symptomatic individuals were tested functionally to find the activity that provoked the symptoms, then taped and retested. No correlation was found between degree of symptom change and radiological change. Roberts concluded that tape may change the pressure concentration on the undersurface of the patella and that these subtle changes may account for the reduction in pain rather than the more observable radiological changes.[67]

McConnell[68] found, using video fluoroscopy of the patella during maximal isometric quadriceps contractions, that taping the patella moved the contact point between the patella and the femur inferiorly. This was suggested to be one possible explanation for the increase in muscle torque

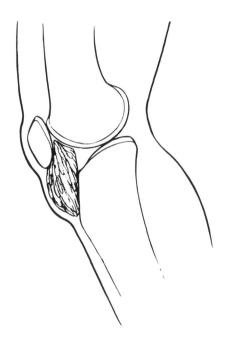

(a)

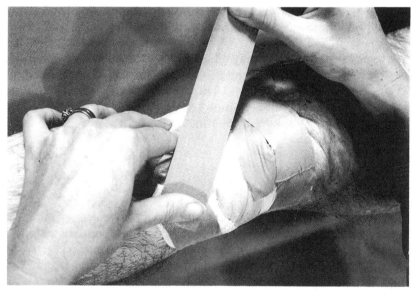

(b)

Figure 7.3 (a) Diagrammatic representation of a posterior tilt of the inferior pole of the patella, irritating the fat pad. (b) An irritated fat pad requires the fat pad to be unloaded. This involves taping superiorly on the patella, to move the inferior pole of the patella anteriorly, as well as, taping from the tibial tubercle to the medial and lateral joint lines, to approximate the attachment of the fat pad.

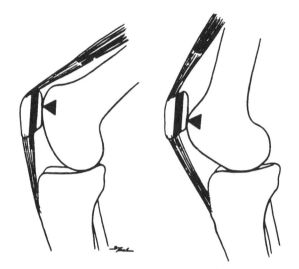

Figure 7.4   Schematic illustrating the concept of the patellofemoral joint acting as a balance beam. The arrow identifies the patellofemoral contact; the striped bar represents the balance beam or lever arm of the patella.[40]

recorded after the patella was taped. Since the patella acts like a balance beam,[37,39,40,69] an inferior position of the contact point will increase the effective lever arm of the quadriceps, especially during extension (Figure 7.4). Therefore, for the same amount of force from the muscle, more torque can be generated.

Tape seems to be necessary only while the VMO, the sole dynamic medial stabilizer of the patellofemoral joint, is being trained. Once trained, the VMO may be actively providing a lengthening of the lateral retinacular structures. The treatment intervention has, therefore, two distinct aspects, symptomatic relief and long-term functionally orientated retraining of muscle action.

Gaining balanced muscle activity

Long-term promotion of optimal joint loading requires analysis of the contribution of each of the surrounding muscles to the mobility and the stability of the joint during various activities, i.e. ensuring balanced muscle activation patterns. Muscle tissue has been shown to have an ideal resting length[63,64] at which it appears to perform optimally with respect to stabilization and generation of force. Elongation of this resting muscle length leads to an increased number of sarcomeres and a decrease in the contractility of the muscle, decreasing the ability of the muscle to exert control over the total range of motion of the segments in question.[70,71]

Although, the contribution of each of the muscles participating in a specific activity can be highly variable among individuals, some patterns of

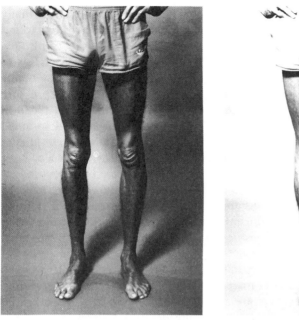

(a)   (b)

Figure 7.5   (a) This man is a competitive distance runner. Over the last 4 months, he has increased the distance and frequency of his training. He is now complaining of bilateral patellofemoral pain. (b) He has marked tibial varum and internal femoral rotation, which are particularly obvious when he puts his feet together.

movement may not be as desirable because they are more likely to cause musculoskeletal problems, particularly when the demand on the segment changes.[1,2,72] An example of undesirable movement patterns, which may predispose certain individuals to musculoskeletal problems, can be demonstrated in subjects with an internally rotated femur (Figure 7.5). These individuals lose flexibility in their tensor fascia latae, which causes a shortening of the iliotibial band, an elongation of the gluteus medius and results in a lateral tracking of the patella, with a consequent increase in length of the VMO.[1,38,59] When these individuals commence an aerobics programme, increase their running routine or rekindle their interest in snow skiing, they are more susceptible to patellofemoral pain.[38,73,74]

Individuals who habitually adopt a forward head posture, develop an elongation of the upper cervical flexors losing the stabilizing ability of these muscles. These individuals are therefore, more susceptible to cervical headaches.[4,75,76] Lack of muscle endurance capacity appears to be of particular importance in the manifestation of cervical headache, and this diminished capacity has been correlated with elongation of the upper cervical muscles.[76]

Physiotherapists need to ensure that the synergistic balance around a joint is restored by giving the patient a specific muscle training programme before the patient commences a generalized strengthening programme, otherwise a poor pattern of recruitment will be reinforced and may actually be harmful.[77–79] For example, patients with low back pain have been found to have greater weakness in their trunk extensors than their trunk flexors.[80–83] Many back-pain patients are routinely given sit-up exercises to increase abdominal strength, yet these exercises may have a potentially harmful effect on these patients. It has been shown that intradiscal pressure in sit-up exercises far exceeds the forces in sitting and equals those in forward bend holding 20 kg.[84,85] If a patient has been advised to avoid bending, lifting or sitting, it does not seem logical to create similar or greater intra-discal pressures during 'therapeutic' sit-up exercises. It has also been suggested that sit-up exercises should be contraindicated for post-menopausal osteoporotic women.[79] In a study on osteoporotic women, subjects who performed through-range abdominal sit-up exercises as a training regimen demonstrated a significantly *increased* incidence of fractures when compared both to a control group and to a group performing isometric extension exercises for the trunk.[79]

It is the lack of endurance capacity of the spinal musculature in patients with back or neck pain, rather than strength,[86,87] which is of greater significance when physiotherapists are considering the type and frequency of training. Specific exercises to promote anterior trunk stabilization and endurance require activation of the internal and external oblique abdominal muscles. This involves either abdominal hollowing and back flattening or abdominal bracing (Figure 7.6). These techniques not only hold the spine in a more neutral position but promote a co-contraction pattern for lumbar spine stability involving both the extensor muscles and the oblique abdominal muscles.[27] A posterior pelvic tilt, on the other hand, increases spinal flexion as well as increasing rectus abdominis activity. Increased rectus abdominis activity has been found to be accompanied by a corresponding decrease in the proportional activation of the erector spinae (e.g. multifidus) and the oblique muscles.[27] Posterior pelvic tilt, according to Richardson and colleagues,[27] should therefore be 'considered more accurately as a movement correction pattern rather than a specific technique to promote lumbar stabilization'. Ballistic trunk movement, such as a rapidly performed sit-up, also increases rectus abdominis activity and decreases the activity of the oblique muscles.[88] This again will adversely affect trunk stabilization.

Specific exercises for stabilizing muscles, whether for the trunk or the limbs must, therefore, be carefully supervised by the therapist, so that appropriate muscles are recruited during the exercise. If there has been habitual disuse of certain muscles, activation will be difficult. Feedback to the patient must be precise to achieve the desired outcome[89] and EMG biofeedback has been found to hasten the rehabilitation process.[90]

In a study comparing EMG biofeedback training of the VMO with resisted short arc quadriceps (SAQ) exercises in an asymptomatic population of young females, Ingersoll and Knight[90] found that although the SAQ group demonstrated an increase in muscle force generation, there was a deterioration of the tracking of the patella during knee motion. The EMG group,

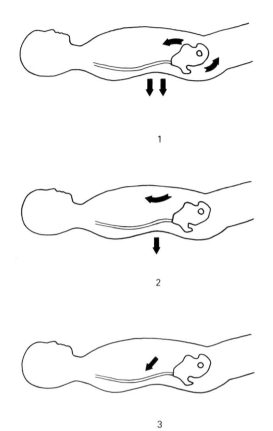

Figure 7.6   Schematic of the pelvic and lumbar spine positions during: (1) posterior pelvic tilt, (2) abdominal hollowing and back flattening, and (3) abdominal bracing.[27]

however, showed an improvement in the congruence angle and 'fit' of the patella in the trochlea. Such findings have significant implications for the rehabilitation of patients with patellofemoral pain and more generally for the rehabilitation of patients with any chronic musculoskeletal problem, where specific training of underused muscles is required to restore the muscle balance around the joint in order to enhance function.

Training muscles specifically requires an understanding of the anatomy and function of the muscle during specific actions. An example of training specificity is given using the VMO. Unlike the 'long' fibres of the vastus medialis, which are active during knee extension, the oblique fibres have no function in extending the knee. In fact, Lieb and Perry[91] found that in cadaver specimens, when weight was applied only to the VMO, the femur fractured before any extension occurred. They found that the vastus medialis oblique was active throughout the entire range of extension and was responsible for realigning the patella medially.

The VMO is the only dynamic medial stabilizer of the knee. The muscle arises from the tendon of the adductor magnus[92] and is supplied, in most cases, by a separate branch of the femoral nerve,[91] so it should be possible to activate it independently of the rest of the quadriceps.[93] To restore the balance between the different quadriceps muscle heads, the therapist should, therefore, emphasize *adduction* of the thigh to the patient in order to facilitate a VMO contraction, rather than extension of the knee. Weight-bearing activities should be commenced early in rehabilitation to improve the synergistic action of muscles linking the pelvis and lower limb segments, which are required for many functional activities of the lower limb. The training needs, therefore, to be specific for the task.[77,78,89] When aiming for muscle strength for a particular activity, the best training is the carefully monitored practice of the activity itself.[89] Strength gains will not be observed when a muscle is required to act in a different functional position to the one where the training has occurred, because different muscle synergies will be required.[77,94] (See also Chapter 6.) As the VMO is a stabilizing muscle,[95,96] endurance training should be emphasized when treating patients with any chronic knee problem. Decreased activity of the VMO has been reported when both asymptomatic control subjects and patellofemoral pain subjects performed rapid ballistic alternating flexion and extension movement of the knee.[95,96] Therefore, exercises should be performed in a slow, controlled fashion until the muscle balance is restored. More rapid activities can be introduced but the muscle activation pattern must be carefully monitored using an EMG biofeedback to ensure that effective segmental alignment is being maintained.

The number of repetitions performed by the patient at a training session will depend upon the onset of muscle fatigue, for example one patient may only be able to step down from a step three times before the quadriceps is quivering, whereas another may be able to maintain a controlled contraction for twenty repetitions. One of the aims of the physiotherapy intervention would be to increase the number of repetitions before the onset of fatigue. Patients should be taught to recognize fatigue so that they do not train through fatigue and risk exacerbating their symptoms.

When prescribing training regimens for adolescents however, the physiotherapist needs to be cognizant of the effect of rapid growth on segmental alignment. Since soft tissue adaptation lags behind growth of long bones, adolescents who have grown rapidly, may demonstrate poor control over body segments and hence be prone to musculoskeletal problems. Their muscles become increasingly inadequate for moving and controlling the body's longer levers. Control over the segments can only be achieved by persistent, specific practice of different activities. The dynamic segmental alignment of these individuals is also affected by soft tissue tightness. Soft tissue tightness occurs in all soft tissues but lack of extensibility of neuro-meningeal tissue is often overlooked as being a potent source of symptoms, especially the plateaued and recurrent symptoms.

Neuromeningeal mobility

Lack of extensibility in the neuromeningeal structures so that the neural tissue inherently does not elongate sufficiently or it becomes adhered to an interface structure can be expected to alter segmental alignment.[98,99] The interface structure may be an anatomical one such as a bone, ligament or muscle, or a pathological interface such as an osteophyte or oedema.[98,99] There is minimal neural elongation relative to the bony interface at certain key locations in the spine, particularly C6, T6 and L4, so the neural tissue often becomes tethered in one or more of these regions. The lack of mobility can be a source of, or contribute to, an individual's symptoms, explaining the increased incidence of remote symptoms related to these locations.

Similar situations are encountered at the posterior knee and the anterior elbow regions where the neural tissue is relatively constrained due to the surrounding soft tissue and is more susceptible to injury.[100] Lack of mobility of the neural tissue will interfere with the performance of smooth controlled movement, especially when the neural tissue is stretched by movement; for example, the sciatic nerve and its branches may become symptomatic when driving a car, or symptoms arising from the median nerve may be experienced when reaching towards the back seat of a car from the driver's seat. Previous injury resulting in inflammation and scarring may mean that the nervous system loses its flexibility and is likely to produce symptoms in a region of the body remote from the original injury.[98,99] For example, a previous Colles' fracture may have caused scar formation around the median nerve. The loss of mobility of the median nerve can lead to restriction in range and pain in the shoulder or the cervical spine and possibly even in the other arm, long after the fracture has healed.[101–104] Symptoms in these regions have often been attributed to either compensatory overuse of the contralateral limb or disuse of the joints and muscles in the symptomatic region. Case study 5 demonstrates the difficulty that can be encountered when determining neuromeningeal involvement.

Case study 5: A 25-year-old beach volleyball player presented with right inferior patella pain, just medial to the patellar tendon. Twelve months previously she had undergone an arthroscopic anterior cruciate ligament reconstruction, using the middle third of her patella tendon. On return to sport, she developed pain in the patella tendon region, so was diagnosed as having patella tendonitis. She rested from sport for 6 months and received physiotherapy treatment localized to the knee joint, with no improvement of symptoms. During the treatment period she had reported some stiffness in the thoracic spine.

On examination, her pain was reproduced on squatting and jumping. She localized the pain to just medial to the patellar tendon. On palpation, distal to the arthroscope portal, there was a thickened, tender area. When the patient was examined in prone, her symptoms were reproduced by flexing the knee and externally rotating the tibia. In this position, palpation of her thoracic spine, specifically over T6 and T7, increased her symptoms. On further questioning, the patient volunteered that the lateral aspect of the knee had a slightly altered sensation.

The symptoms in this case were probably due to a scarring of the infrapatellar branch of the saphenous nerve. The cutaneous nerve distribution of

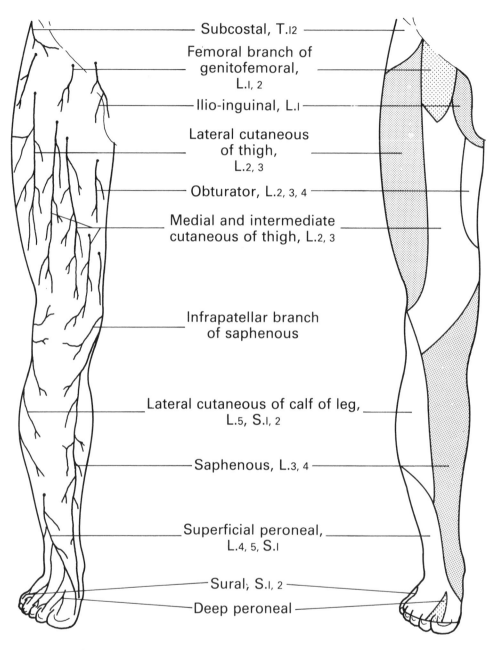

Subcostal, T.12

Femoral branch of genitofemoral, L.1, 2

Ilio-inguinal, L.1

Lateral cutaneous of thigh, L.2, 3

Obturator, L.2, 3, 4

Medial and intermediate cutaneous of thigh, L.2, 3

Infrapatellar branch of saphenous

Lateral cutaneous of calf of leg, L.5, S.1, 2

Saphenous, L.3, 4

Superficial peroneal, L.4, 5, S.1

Sural, S.1, 2

Deep peroneal

Figure 7.7   The cutaneous nerves of the right lower limb, viewed from the anterior aspect.[105]

the lower leg is illustrated in Figure 7.7. Although the saphenous nerve is derived from the lumbar plexus, the meninges are continuous over the length of the spine. The proximal level at which tethering may have occurred was T6, hence the associated symptoms.

Treatment was aimed at freeing the nerve. This was done locally at the site of thickening, by flexing the knee, externally rotating the tibia and stretching the soft tissue around the medial side of the knee. Treatment was also directed to the site of tethering in the spine by mobilizing the thoracic spine while the peripheral nerve was in the stretched position. The short-term results of this intervention were an immediate reduction in painful symptoms. The patient was given a home stretching programme and a muscle training programme for the VMO and gluteals, as these muscles exhibited suboptimal control. This programme maintained the symptom-free state and allowed the patient to return quickly to competitive sport.

CONCLUSION

Effective management of musculoskeletal problems involves comprehensive problem analysis, an understanding of the type of environment in which the muscles will be acting and a specific management programme to ensure the desired changes occur. As with all training, success or failure depends on the amount and quality of practice. The training must be simple, requiring minimal equipment so that it is readily accessible to individuals and can be repeated frequently.

Promotion of an effective segmental alignment may prevent the onset or recurrence of musculoskeletal problems. This alignment will be achieved through improved joint congruity, soft tissue flexibility and dynamic balance between the muscles controlling the segments involved in the movement. The physiotherapist can only help the patient to achieve this through a thorough understanding of the relevant anatomy, physiology, neurophysiology and biomechanical aspects of the problem and an awareness of behavioural influences on the symptoms. As physiotherapists increase their knowledge in the movement sciences, they should be at the 'cutting edge', evaluating and continually improving their therapeutic intervention.

REFERENCES

1. Sahrmann S.A. (1990). The movement system balance theory: relationship to musculoskeletal pain syndromes. Unpublished manuscript.
2. Sahrmann S.A. (1987). Muscle imbalances in the orthopaedic and neurological patient. *Proceedings of the Tenth International Congress of the World Confederation for Physical Therapy*, Sydney, New South Wales, Australia, p. 836.
3. Kendall H.D., Kendall F.P., Boynton D.A. (1952). *Posture and Pain*. Baltimore: Williams and Wilkins.
4. Janda V. (1988). Muscle and cervicogenic pain syndromes. In *Physical Therapy of the Cervical and Thoracic Spine* (Grant E.R. ed.). New York: Churchill Livingstone.
5. Rocabado M. (1984). Biomechanical analysis of the craniocervical relationships through teleradiographic tracings. *Proceedings of 5th International Conference, International Federation of Orthopaedic Manipulative Therapists*, Vancouver, Canada, p. 14.
6. Asher C. (1975). *Postural Variations in Childhood*. London: Butterworth.

7. Magee D.J. (1987). *Orthopaedic Physical Assessment*. Philadelphia: WB Saunders.
8. During J., Goudfrooij H., Keessen W., *et al.* (1985). Towards standards for posture. Postural characteristics of the lower back system in normal and pathological conditions. *Spine*, **10**, 88.
9. Andersson G.B.T., Ortengen R., Nachemson A. (1977). Intradiscal pressure, intra-abdominal pressure and myoelectric back activity related to posture and loading. *Clin. Orthop.*, **129**, 156.
10. Dolan P., Adams M., Hutton W. (1988). Commonly adopted postures and their effect on the lumbar spine. *Spine*, **13**, 197.
11. Adams M., Hutton W. (1985). The effect of posture on the lumbar spine. *J. Bone Joint Surg.*, **67B**, 635.
12. Aspden R.M. (1988). A new mathematical model of the spine and its application of spinal loading in the workplace. *Appl. Ergonomics*, **19**, 319.
13. Gracovetsky S., Farfan H.F., Lamy C. (1977). A mathematical model of the lumbar spine using an optimal system to control muscles and ligaments. *Orthop. Clin. North Am.*, **8**, 135.
14. Cross M.J., Crichton K.J. (1987). *Clinical Examination of the Injured Knee*. London: Harper and Row.
15. Meillon M.B., Walsh W.M., Shelton G.L. (1990). *The Team Physician's Handbook*. Philadelphia: Hanley and Belfus.
16. Frankel V.H., Nordin M. (1980). *Basic Biomechanics of the Skeletal System*. Philadelphia: Lea and Febiger.
17. Kapandji I.A. (1970). *The Physiology of Joints*. Vol. 1. New York: Churchill Livingstone.
18. Bowling R.W., Rockar P.A., Erhard R. (1986). Examination of the shoulder complex. *Phys. Ther.*, **66**, 1866.
19. Hollingshead W.H., Jenkins D.B. (1981). *Functional Anatomy of the Limbs and Back*. Philadelphia: WB Saunders.
20. Ayub E. (1987). Posture and the upper quarter. In *Physical Therapy of the Shoulder* (Donatelli R. ed.). New York: Churchill Livingstone.
21. Perry J. (1983). Shoulder anatomy and biomechanics. *Clin. Sports Med.*, **2**, 247.
22. Norkin C., Levangie P. (1983). *Joint Structure and Function*. Philadelphia: FA Davis.
23. Crawford H.J., Jull G.A. (1991). The influence of thoracic form and movement on range of shoulder flexion. *Proceedings MPAA 7th Biennial Conference*. New South Wales, Australia, p. 154.
24. Rothstein J.M. (1985). Measurement and clinical practice: theory and application. In *Measurement in Physical Therapy* (Rothstein J.M. ed.). New York: Churchill Livingstone.
25. Payton O.D. (1988). *Research: The Validation of Clinical Practice*. 2nd edn. Philadelphia: FA Davis.
26. Zussman M. (1986). The absolute visual analogue scale as a measure of pain intensity. *Aust. J. Physiother.*, **32**, 244.
27. Richardson C., Jull G., Toppenberg R., *et al.* (1992). Techniques for active lumbar stabilisation for spinal protection: a pilot study. *Aust. J. Physiother.*, **38**, 105.
28. Turner-Smith A.R., Harris D. (1985). Shape measurement in the scoliosis clinic. In *Biomechanical Measurement in Orthopaedic Practice* (Whittle M., Harris D. eds.). Oxford: Butterworth-Heinemann, p. 92.
29. Winter D.A. (1982). Camera speeds for normal and pathological gait analysis. *Med. Biol. Eng. Comput.*, **20**, 407.
30. Gracovetsky S., Newman N., Asselin S. (1990). *The Problem of Non-Invasive Assessment of Spinal Function*. In-house publication.
31. D'Angelo M.D., Greive D.W., Pereira L.F., *et al.* (1987). A description of normal relaxed standing postures. *Clin. Biomech.*, **2**, 140.

32. Ada L., Westwood P. (1992). A kinematic analysis of recovery of the ability to stand up following stroke. *Aust. J. Physiother.*, **38**, 135.
33. Helminen H., Kiviranta I., Tammi M., *et al.* (1987). *Joint Loading*. London: Butterworth.
34. Brandt K. (1981). Pathogenesis of osteoarthritis. In *Textbook of Rheumatology* (Kelley W.N., Harris E.D., Ruddy S., Sledge C.B. eds.). Philadelphia: WB Saunders.
35. Mow V., Eisenfeld J., Redler I. (1974). Some surface characteristics of articular cartilage II: On the stability of articular surface and a possible biomechanical factor in aetiology of chondrodegeneration. *J. Biomech.*, **7**, 457.
36. Merriam J.L. (1980). *Engineering Mechanics Volume 1: Statics*. New York: John Wiley and Sons.
37. Hungerford D.S., Barry M. (1979). Biomechanics of the patello-femoral joint. *Clin. Orthopa.*, **144**, 9.
38. Fulkerson J., Hungerford D. (1990). *Disorders of the Patellofemoral Joint*. 2nd edn. Baltimore: Williams & Wilkins.
39. Yamaguchi G., Zajac F. (1989). A planar model of the knee joint to characterize the knee extensor mechanism. *J. Biomech.*, **22**, 1.
40. Buff H., Jones L.C., Hungerford D.S. (1988). Experimental determination of forces transmitted through the patellofemoral joint. *J. Biomech.*, **21**, 17.
41. Ahmed A.M., Burke D.L., Yu A. (1983). *In-vitro* measurement of static pressure distribution in synovial joints – II. Retropatellar surface. *J. Biomed. Eng.*, **105**, 226.
42. Huberti H.H., Hayes W.C., Stone J.L., *et al.* (1984). Force ratios in the quadriceps tendon and ligamentum patellae. *J. Orthop. Res.*, **2**, 49.
43. Huberti H., Hayes W. (1984). Patellofemoral contact pressures. *J. Bone Joint Surg.*, **66A**, 715.
44. Matthews L., Sonstegard D., Henke J. (1977). Load bearing characteristics of the patellofemoral joint. *Acta Orthop. Scand.*, **48**, 511.
45. Reilly D. Martens M. (1972). Experimental analyses of the quadriceps muscle force and patellofemoral joint reaction force for various activities. *Acta Orthop. Scand.*, **43**, 126.
46. Goodfellow J., Hungerford D., Zindel M. (1976). Patellofemoral joint mechanics and pathology. *J. Bone Joint Surg.*, **58B**, 287.
47. Dahhan P., Delphine G., Larde D. (1981). The femoropatellar joint. *Anat. Clin.*, **3**, 23.
48. Fujikawa K., Seedholm B., Wright V. (1983). Biomechanics of the patellofemoral joint: 1. A study of the patellofemoral compartment and movement of the patella. *Eng. Med.*, **12**, 3.
49. Radin E. (1979). A rational approach to treatment of patellofemoral pain. *Clin. Orthop.*, **144**, 107.
50. Kramer P.G. (1986). Patella malalignment syndrome: rationale to reduce excessive lateral pressure. *J. Orthop. Sports Phys. Ther.*, **8**, 301.
51. Fox T.A. (1975). Dysplasia of the quadriceps mechanism. *Surg. Clin. North Am.*, **55**, 199.
52. Root M., Orien W., Weed J. (1977). *Clinical Biomechanics* Vol. II. Los Angeles: Clinical Biomechanics.
53. Seedholm B., Takeda T., Tsubuku M., *et al.* (1979). Mechanical factors and patellofemoral osteoarthritis. *Ann. Rheum. Dis.*, **38**, 307.
54. Speakman G.B., Weisberg J. (1977). The vastus medialis controversy. *Physiotherapy*, **63**, 249.
55. Spencer J., Hayes K., Alexander I. (1984). Knee joint effusion and quadriceps reflex inhibition in man. *Arch. Phys. Med. Rehabil.*, **65**, 171.
56. Stokes M., Young A. (1984). Investigations of quadriceps inhibition: implications for clinical practice. *Physiotherapy*, **70**, 425.

57. Subotnik S. (1980). The foot and sports medicine. *J. Orthop. Sports Phys. Ther.*, **2**, 53.
58. Tiberio D. (1987). The effect of excessive subtalar joint pronation on patello-femoral mechanics; a theoretical model. *J. Orthop. Sports Phys. Ther.*, **9**, 160.
59. McConnell J. (1983). *An investigation of certain biomechanical variables predisposing an adolescent male to retropatellar pain.* Unpublished project report, Graduate Diploma in Manipulative Therapy, Sydney, Australia.
60. McConnell J. (1986). The management of chondromalacia patellae – a long term solution. *Aust. J. Physiother.*, **32**, 215.
61. McConnell J. (1987). Training the vastus medialis oblique in the management of patellofemoral pain. *Proceedings of Tenth Congress of the World Confederation for Physical Therapy*. Sydney, Australia.
62. McConnell J. (1987). Patellar alignment and quadriceps strength. *Proceedings of 5th Biennial Conference, Manipulative Therapists Association of Australia*. Melbourne, Australia.
63. Leveau B.F. (1985). Basic biomechanics in sports and orthopaedic therapy. In *Orthopaedic and Sports Physical Therapy* (Gould J. A., Davies G.J. eds.). St Louis: CV Mosby.
64. Dumbleton J.H., Black J. (1975). *An Introduction to Orthopaedic Materials.* Springfield: Charles C. Thomas.
65. White A.A., Panjabi M.M. (1978). *Clinical Biomechanics of the Spine.* Philadelphia: JB Lippincott.
66. McConnell J. (1991). Fat pad irritation – a mistaken patellar tendonitis. *Sport Health*, **9**, 7.
67. Roberts J.M. (1989). The effect of taping on patellofemoral alignment – a radiological pilot study. *Proceedings MTAA 6th Biennial Conference*, Adelaide, South Australia.
68. McConnell J.S. (1992). *A mechanical investigation into the effect of taping the patella of patients with patellofemoral pain.* Unpublished Masters Thesis.
69. Maquet P. (1984). *Biomechanics of the Knee.* Berlin: Springer-Verlag.
70. Gossman M., Sahrmann S.A., Rose S.J. (1982). Review of length-associated changes in muscle. Experimental and clinical implications. *Phys. Ther.*, **62**, 1799.
71. Williams P.E., Goldspink G. (1978). Changes in sarcomere length and physiological properties in immobilised muscle. *J. Anat.*, **127**, 459.
72. Kendall H.O., Kendall F.P. (1968). Developing and maintaining good posture. *Phys. Ther.*, **48**, 319.
73. Malek M., Mangine R. (1981). Patellofemoral pain syndromes: a comprehensive and conservative approach. *J. Orthop. Sports Phys. Ther.*, **2**, 108.
74. Pevsner D., Johnson A., Blazina G. (1979). The patellofemoral joint and its implications in the rehabilitation of the knee. *Phys. Ther.*, **57**, 869.
75. Jull G.A. (1986). Headaches of cervical origin. In *Physical Therapy of the Cervical and Thoracic Spine. Clinics in Physical Therapy* (Grant E.R. ed.). New York: Churchill Livingstone.
76. Watson D.H., Trott P.H. (1991). Cervical headache: an investigation of natural head posture and upper cervical flexor muscle performance. *Proceedings MPAA 7th Biennial Conference*. New South Wales, Australia.
77. Sale D., MacDougall D. (1981). Specificity of strength training: a review for coach and athlete. *Can. J. Appl. Sports Sci.*, **6**, 87.
78. Sale D. (1987). Influence of exercise and training on motor unit activation. *Exerc. Sports Sci. Rev.*, **5**, 95.
79. Sinaki M., Mikkelson B.A. (1984). Post menopausal osteoporosis: flexion versus extension exercises. *Arch. Phys. Med. Rehabil.*, **65**, 593.
80. Addison R., Schultz A. (1980). Trunk strengths in patients seeking hospitalisation for chronic low back disorders. *Spine*, **5**, 539.

81. McNeil T., Warwick D., Andersson G., et al. (1980). Trunk strengths in attempted flexion, extension and lateral bending in healthy subjects and patients with low back disorders. *Spine*, **5**, 529.
82. Mayer T.G., Gatchel R.J., Kishino N., et al. (1986). A prospective short term study of chronic low back pain patients utilizing novel objective function measurement. *Pain*, **25**, 53.
83. Pope M.H., Bevins T., Wilder D.G., et al. (1985). The relationship between anthropometric, postural, muscular and mobility characteristics of males aged 18–55. *Spine*, **10**, 644.
84. Nachemson A.L. (1981). Disc pressure measurement. *Spine*, **6**, 93.
85. Nachemson A.L., Elfstrom G. (1970). Intravital dynamic pressure measurements in lumbar discs. A study of common movements, manoeuvres and exercises. *Scand. J. Rehabil. Med. Suppl.*, **1**, 1.
86. Biering-Sorensen F. (1984). Physical measurements as risk indicators of low back trouble over a one year period. *Spine*, **9**, 106.
87. Suzuki N., Endo S. (1983). A quantitative study of trunk muscle strength and fatiguability in the low back pain syndrome. *Spine*, **8**, 69.
88. Richardson C., Jull G., Wohlfahrt D. (1991). Ballistic exercise: can it undermine the protective stability role of the lumbar musculature? *Proceedings MPAA 7th Biennial Conference*. New South Wales, Australia.
89. Carr J., Shepherd R. (1982). *A Motor Relearning Programme for Stroke*. London: Heinemann.
90. Ingersoll C., Knight K. (1991). Patellar location changes following EMG biofeedback or progressive resistive exercises. *Med. Sci. Sports Exerc.*, **23**, 1122.
91. Lieb F., Perry J. (1968). Quadriceps function. *J. Bone Joint Surg.*, **50A**, 1535.
92. Bose K., Kanagasuntherum R., Osman M. (1980). Vastus medialis oblique: an anatomical and physiologic study. *Orthopedics*, **3**, 880.
93. Basmajian J.V., De Luca C.J. (1985). *Muscles Alive*. Baltimore: Williams & Wilkins.
94. Astrand P., Rodahl K. (1982). *Textbook of Work Physiology*. New York: McGraw Hill.
95. Richardson C.A., Bullock M.I. (1986). Changes in muscle activity during fast alternating flexion and extension movements of the knee. *Scand. J. Rehabil. Med.*, **18**, 51.
96. Richardson C. (1987). Atrophy of vastus medialis in patellofemoral pain syndrome. *Proceedings of the 10th International Congress of the World Confederation of Physical Therapy*. Sydney, Australia, p. 400.
97. Micheli L., Slater J., Woods E., et al. (1986). Patella alta and the adolescent growth spurt. *Clin. Orthop.*, **213**, 159.
98. Breig A. (1978). *Abnormal Mechanical Tension in the Central Nervous System*. Stockholm: Almqvist & Wiksell.
99. Butler D.S. (1991). *Mobilisation of the Nervous System*. London: Churchill Livingstone.
100. Louis R. (1981). Vertebroradicular and vertebromedullar dynamics. *Anat. Clin.*, **3**, 1.
101. Aro H., Koivenun T., Katevuo K., et al. (1988). Late compression neuropathies after Colles' fracture. *Clin. Orthop.*, **233**, 217.
102. Cooney W.P., Dobyns J.H., Linscheid R.L. (1980). Complications of Colles' fractures. *J. Bone Joint Surg.*, **62A**, 613.
103. Frykman G. (1967). Fracture of the distal radius including sequellae: shoulder–hand–finger syndrome, disturbance in the distal radio-ulnar joint and impairment of nerve function: a clinical and experimental study. *Acta Orthop. Scand. Suppl.*, **108**, 1.

104. *Stewart H.O., Innes A.R., Burke F.D. (1985). The hand complications of Colles' fracture. J. Hand Surg.,* **10B**, 103.
105. Williams P., Warwick R. (1980). *Gray's Anatomy.* London: Churchill Livingstone.

Index